Other Titles in This Series

8 **Winfried Just and Martin Weese,** Discovering modern set theory. I: The basics, 1996

7 **Gerald J. Janusz,** Algebraic number fields, second edition, 1996

6 **Jens Carsten Jantzen,** Lectures on quantum groups, 1996

5 **Rick Miranda,** Algebraic curves and Riemann surfaces, 1995

4 **Russell A. Gordon,** The integrals of Lebesgue, Denjoy, Perron, and Henstock, 1994

3 **William W. Adams and Philippe Loustaunau,** An introduction to Gröbner bases, 1994

2 **Jack Graver, Brigitte Servatius, and Herman Servatius,** Combinatorial rigidity, 1993

1 **Ethan Akin,** The general topology of dynamical systems, 1993

Discovering Modern Set Theory. I

GRADUATE STUDIES IN MATHEMATICS 8

Discovering Modern Set Theory. I

Winfried Just
Martin Weese

AMERICAN MATHEMATICAL SOCIETY
Providence, Rhode Island

2000 *Mathematics Subject Classification.* Primary 03Exx;
Secondary 03E30, 03E25, 03E10, 03E05.

ABSTRACT. This volume is an introduction to set theory for beginning graduate students. It covers the basics of set theory that are considered prerequisites for other areas of mathematics such as ordinals, cardinals, transfinite induction and recursion, and applications of Zorn's lemma. It also contains a description of how mathematics can be founded on axiomatic set theory, as well as a discussion of the nature of consistency results.

Library of Congress Cataloging-in-Publication Data

Just, W. (Winfried)
Discovering modern set theory / Winfried Just, Martin Weese.
p. cm. — (Graduate studies in mathematics, ISSN 1065-7339; v. 8)
Includes bibliographical references and index.
Contents: 1. The basics
ISBN 0-8218-0266-6 (v. 1 : hard cover : alk. paper)
1. Set theory. I. Weese, Martin. II. Title. III. Series.

QA248.J87 1995 95-44663
511.3′22—dc20 CIP
Softcover ISBN 978-1-4704-8117-9

∞ The paper used in this book is acid-free and falls within the guidelines
established to ensure permanence and durability.
Visit the AMS home page at `http://www.ams.org/`

12 11 10 9 8 7 6 5 30 29 28 27 26 25

To Ewelina and Veronika

Contents

Preface

This text grew out of lecture notes written[1] by Martin Weese of Humboldt Universität, Berlin, Germany in 1992/93. Winfried Just used part of these notes during the same academic year in a lecture course on set theory for first and second year graduate students at Ohio University, Athens, Ohio. While doing this, the idea of writing up an English text of the lecture crossed his mind. The idea proved irresistible enough to result in this book.

The genesis of this text accounts for some departures from the traditional textbook format. Generally speaking, this is not so much a textbook that could form the backbone of a lecture, but rather the text of the lecture itself. Accordingly, our language is perhaps closer to spoken, colloquial English than to the standard style of mathematical writing. Frequently, the text takes on the form of a dialogue between the authors and the reader. The exercises form an integral part of this dialogue and are not relegated to the end of sections.

We tried to keep the length of the text moderate. This may explain the absence of many a worthy theorem from this book. Our most important criterion for inclusion of an item was frequency of use outside of pure set theory. We want to emphasize that "item" may mean either an important concept (like "equiconsistency with the existence of a measurable cardinal"), a theorem (like Ramsey's Theorem), or a proof technique (like the craft of using Martin's Axiom). Therefore, we occasionally illustrate a technique by proving a somewhat marginal theorem. Of course, the "frequency of use outside set theory" is based on our subjective perceptions.

We do feel some remorse for the total exclusion of descriptive set theory from this text. This is a very important branch of set theory, and it overlaps with several other areas of mathematics, most notably topology and recursion theory. However, just adding a section on descriptive set theory to a text like this would look artificial and do no justice to the area. It seems to us that one should either cover a lot of descriptive set theory, or none at all.

At the end of most sections, there are "Mathographical Remarks." Their purpose is to show where the material fits in the history and literature of the subject. We hope they will provide some guidance for further reading in set theory. They should not be mistaken for "scholarly remarks" though. We did not make any effort whatsoever to trace the theorems of this book to their origins. However, each of the theorems presented here can also be found in at least one of the more specialized texts reviewed in the "Mathographical Remarks." Therefore, we do not feel guilty of severing chains of historical evidence.

This book owes its existence as much to our students and colleagues as it does to its authors. Parts of the original version of Martin Weese's lecture notes were

[1] In German.

read by students of Humboldt University, Berlin. We thank them for pointing out mistakes and suggesting improvements.

We are also indebted to Jörg Brendle for many thoughtful comments on the German version.

Special thanks are due to Mary Anne Swardson of Ohio University who read the very first English version of this text and generously applied her red pencil to it.[2] We are much indebted to her for this invaluable service.

We thank Howard Wicke of Ohio University, Marion Scheepers of Boise State University, and Frank Tall of the University of Toronto for reading parts of later versions and commenting on their shortcomings. Last but not least, we thank Ohio University students Brian Johnson, Mark McKibben, Todd Allin Morman, William Stamp, and Mark Starr for struggling through parts of this book and making many valuable suggestions for improvement.

[2] As the old saying goes: "Spare the red and spoil the text."

How to Read this Book

So we shall now explain how to read the book. The right way is to put it on your desk in the day, below your pillow at night, devoting yourself to the reading, and solving the exercises till you know it by heart. Unfortunately, I suspect the reader is looking for advice how not to read, i.e. what to skip, and even better, how to read only some isolated highlights.

(Saharon Shelah in the introduction to his book "Classification Theory and the Number of Non-Isomorphic Models")

In mathematics, as anywhere today, it is becoming more difficult to tell the truth. To be sure, our store of accurate facts is more plentiful now than it has ever been. ... Unfortunately, telling the truth is not quite the same as reciting a rosary of facts.

(Gian-Carlo Rota, 1985)

W. A. Hurwitz used to say that in teaching on an elementary level one must tell the truth, nothing but the truth, but not the whole *truth.*

(Mark Kac, 1976)

We wrote this text for two kinds of readers: beginning graduate students who want to get some grounding in set theory, and more advanced mathematicians who wish to broaden their knowledge of set theory. Furthermore, we wanted this text to be useful both as a textbook for a regular graduate course, as well as for those readers who wish to use it without the guidance of an instructor.

Volume I contains the basics of modern set theory. Many graduate texts on analysis, algebra, topology, or measure theory begin with a review of parts of this material as "set-theoretic prerequisites." Thus, Volume I is primarily aimed at beginning graduate or advanced undergraduate students. It can be used as a textbook in an introductory set theory course, or as supplementary reading in a course that relies heavily on set-theoretic prerequisites. Volume II is aimed at more advanced graduate students and research mathematicians specializing in fields other than set theory. It contains short but rigorous introductions to various set-theoretic techniques that have found applications outside of set theory. Although we think of Volume II as a natural continuation of Volume I, each volume is sufficiently self-contained to be studied separately. Since our terminology is fairly standard, more advanced students may be able to skip the first few sections of Volume I or even go directly to Volume II.

If you do not have the benefit of an instructor who can tell you what to skip, the best policy is to proceed as follows: Read the Introduction. It will give you some general idea what we are up to. If you don't understand every word of it, don't worry. Next, find out at which point of the rest of the book things start to look new to you, and begin reading right there. If things start to look new only around Chapter 22 or so, you probably do not want to waste your time reading this book, but go to the more advanced literature on the subject. In the "mathographical remarks" at the end of most chapters, you will find ample suggestions for further reading.

If things look new to you right from the beginning, check whether you know most of the concepts and symbols listed under "Basic Notations". If so, read Chapter 1, where some of the prerequisite material is reviewed. If Chapter 1 is pleasant, easy reading, then you are probably ready for this book. If more than two concepts listed under "Basic Notations" are entirely new to you, or if Chapter 1 feels challenging, then you may want to read one of the more elementary texts listed in the mathographical remarks at the end of Chapter 1.

Roughly speaking, the only prerequisite for this book is that you are at ease with set-theoretic notation. However, some knowledge of mathematical logic and (for Volume II) general topology is indispensible. Therefore, to make the exposition somewhat self-contained, we included a minicourse in mathematical logic in Chapters 5 and 6, and also an Appendix on general topology at the end of Volume II.

Once you have determined your point of entrance, it is best to read the rest of the book line by line. Much of this book is written like a dialogue between the authors and the reader. This is intended to model the practice of creative mathematical thinking, which more often than not takes on the form of an inner dialogue in a mathematician's mind. You will quickly notice that this text contains many question marks. This reflects our conviction that in the mathematical thought process it is at least as important to have a knack for asking the right questions at the right time as it is to know some of the answers.

You will benefit from this format only if you do your part and actively participate in the dialogue. This means in particular: Whenever we pose a rhetorical question, pause for a moment and ponder the question before you read our answer. Sometimes we put a little more pressure on you and call our rhetorical questions EXERCISES. Not all exercises are rhetorical questions that will be answered a few lines later. Sometimes, the completion of a proof is left as an exercise. We also may ask you to supply the entire proof of an interesting theorem, or an important example. Nevertheless, we recommend that you attempt the exercises right away, especially all the easier ones. Most of the time it will be easier to digest the ensuing text if you have worked on the exercise, even if you were unable to solve it.

Here is a well-kept secret: All mathematical research papers contain plenty of exercises. These usually appear under the disguise of seemingly unnecessary assumptions, missing examples, or phrases like: "It is easy to see." One of the most important steps in becoming a mathematician is to learn to recognize hidden EXERCISES, and to develop the habit of tackling them right away.

We often make references to solutions of exercises from earlier chapters. Sometimes, the new material will make an old and originally quite hard exercise seem trivial, and sometimes a new question can be answered by modifying the solution to a previous problem. Therefore, it is a good idea to collect your solutions and even your failed attempts at solutions in a folder where you can look them up later.

The level of difficulty of our exercises varies greatly. To help the reader save time, we rated each exercise according to what we perceive as its level of difficulty. The rating system is the same as used by American movie theatres. Everybody should attempt the exercises rated G (general audience). Beginners are encouraged to also attempt exercises rated PG (parental guidance), but may sometimes want to consult their instructor for a hint. It is also a good idea to double–check your solution with the instructor, especially if it looks trivial to you. Exercises rated R (restricted) are intended for mature audiences. The X–rated problems must not be attempted by anyone easily offended or discouraged.

There is another important reason why we do not recommend skipping chapters. Mathematical formalism is a good thing, but it is secondary to the development of the ideas that are being formalised. In the spirit of A. W. Hurwitz, we shall introduce many of the more difficult concepts in stages: first intuitively, perhaps by a suggestive analogy, and later in the text with full mathematical rigor. By skipping ahead, you may miss the more rigorous treatments of a concept and become stuck with some vague intuitive notions that you should have long outgrown. The latter problem may be alleviated by making good use of the index. Also, a footnote often alerts the reader when telling the whole truth is postponed.

If Theorem 4 of Chapter 17 is referred to in Chapter 17 itself, it will be called just Theorem 4. Outside of Chapter 17, it will be called Theorem 17.4.

The end of a proof is usually marked by a $\square$.

Mathographical Remark

The quotes of Rota and Kac are taken from *Discrete thoughts*, by Mark Kac, Gian-Carlo Rota and Jacob T. Schwartz, Birkhäuser, Boston 1992, pages ix and 15.

Basic Notations

We assume that you are familiar with the concepts represented by the following symbols.
$\{x_0, x_1, \ldots, x_n\}$ — the set containing $x_0, x_1, \ldots, x_n$ and no other elements;
$\emptyset$ — the empty set;
$\in$ — the membership relation;
$\subseteq$ — subset;
$\subset$ — proper subset;
$\exists, \forall, \neg, \wedge, \vee, \rightarrow, \leftrightarrow$ — quantifiers and logical connectives. Although Chapter 5 contains a thorough discussion of these symbols in the context of formal languages, you should already be at ease with their use.
$x \cup y = \{z : z \in x \vee z \in y\}$ — the union of two sets;
$x \cap y = \{z : z \in x \wedge z \in y\}$ — the intersection of two sets;
$x \backslash y = \{z : z \in x \wedge z \notin y\}$ — the difference of two sets;
$x \Delta y = (x \backslash y) \cup (y \backslash x) = (x \cup y) \backslash (x \cap y)$ — the symmetric difference of two sets;
$\bigcup \mathcal{X} = \{z : \exists Y \in \mathcal{X}\, (z \in Y)\}$ — union of a family of sets;
$\bigcap \mathcal{X} = \{z : \forall Y \in \mathcal{X}\, (z \in Y)\}$ — intersection of a family of sets;
$f : X \rightarrow Y$ — function from X into Y;
$f[W] = \{y \in Y : \exists x \in W\, f(x) = y\}$ — image of W under f;
$f^{-1}Z = \{x \in X : f(x) \in Z\}$ — inverse image of Z under f;
$dom(f)$ — domain of a function f;
$rng(f)$ — range of a function f;
$f|W$ — restriction of a function f to a subset W of its domain;
$f \circ g$ — composition of two functions;
$\langle a_n : n \in \mathbb{N} \rangle = (a_n)_{n \in \mathbb{N}}$ — sequence indexed by natural numbers;
$\mathbb{N} = \{0, 1, 2, \ldots\}$ — the set of natural numbers;
$\mathbb{Z}$ — the set of integers;
$\mathbb{Q}$ — the set of rationals;
$\mathbb{R}$ — the set of reals;
$\mathbb{P} = \mathbb{R} \backslash \mathbb{Q}$ — the set of irrationals;
$\mathbb{C}$ — the set of complex numbers.
$\mathbb{A}$ — the set of algebraic numbers

Note that we impose no restrictions on the style of letters that represent sets. Each of the symbols $x, X, \mathcal{X}, \mathbb{X}$ may stand for a set.

Introduction

No one shall be able to drive us from the paradise that Cantor created for us.
David Hilbert, 1925

Whatever the remedy adopted (for resolving the paradoxes of set theory), *we can promise ourselves the joy of a doctor called to witness an interesting pathological case.*
Henri Poincaré, 1908

In 1873, the German mathematician Georg Cantor discovered that the set of algebraic reals is countable. A few weeks later he was able to demonstrate that the set of all real numbers is uncountable. A new mathematical discipline was born: set theory. In the course of the next two decades, Cantor developed the fundamental concepts of this new discipline; the concepts of equipotent sets, order-isomorphic structures, cardinals and ordinals are all due to him.

Generally speaking, set theory is the study of collections of objects. This view was expressed by Cantor in his famous definition of a set:

> *By a "set" we mean any collection M into a whole of definite, distinct objects m (which are called the "elements" of M) of our perception or of our thought.*

If an element m belongs to a set M, we write $m \in M$. It is also quite common to say in this case that *m is a member of M*, and to refer to $\in$ as the *membership relation*. As we shall see, all relevant facts about sets can be expressed in terms of the membership relation.

The career of set theory has been impressive. In the first half of the 20th century, the new fields of set-theoretic topology, theory of real functions, and functional analysis evolved. Each of these disciplines is strongly rooted in set theory, albeit not exclusively. Even more importantly, set theory can be regarded as the foundation of all mathematics. It is possible to interpret the other branches of mathematics as the study of sets. This seems at first glance to be an implausible claim. For most people the real number $\sqrt{2}$ is just a single object, perhaps a point on the real line, but certainly not a collection of other objects.

EXERCISE 1(X): If you already know some methods of reinterpreting mathematics in terms of set theory, use it to rewrite the Fundamental Theorem of Calculus in the language of set theory.

Nevertheless, in the sections ahead, we shall present compelling evidence for the possibility of founding all of mathematics on set theory.

But why bother? Many mathematicians, especially those of the more applied persuasion, never use even such basic set-theoretic tools as arithmetic of infinite ordinals in their research. Would it not be more reasonable to study set theory just as a separate discipline, rather than trying to fit all other disciplines into a set-theoretic straightjacket? Or, if the topologists really cannot live without the straightjacket, shouldn't at least the applied areas be spared?

This suggestion misses the point on two counts. First, it is one thing to claim that set theory could *in principle* serve as a foundation for all of mathematics, including, say, differential equations; and it is quite another thing to seriously propose that the Navier-Stokes equations should be expressed in the language of set theory. If you tried your luck with Exercise 1, you know that the latter is an absurdity which nobody in his right mind would seriously consider. Second, there is a definite advantage to having a single framework for the separate subdisciplines of mathematics. Different branches of mathematics build on each other. Analytic functions are heavily used in number theory, for example. If each branch of mathematics had its own separate foundations, each use of a theorem from another subfield might raise foundational issues. The existence of a common, albeit at times clumsy, framework makes such mathematical cross-breeding entirely unproblematic. It is exactly the existence of the established common framework that allows most practioners to just do mathematics and to leave all foundational issues to the specialists: set theorists, logicians and philosophers.

The suggestion "to spare at least the applied areas from the straightjacket of set theory" may sound funny, but it expresses a belief that is held in earnest by many mathematicians and science administrators: that one can draw a clear dividing line between applied and pure mathematics. Even those who do not share this view tend to think that there are clear-cut instances of belonging to the realm of either the pure or the applied. And yet, the distance between the most applied and the most abstract may be surprisingly short. Consider probability theory. This is as applied a field as any. To a mathematician, it is just the study of probability measures. These are functions defined on certain σ–fields of sets, for instance on the Lebesgue measurable subsets of the unit interval. Once this framework for doing probability theory has been established, it is very natural to ask whether there exists a probability function defined on *all* subsets of the unit interval so that, as in the case of the familiar Lebesgue measure, each individual point has probability zero. As we shall see in Chapter 23, this is one of the deepest and most perplexing problems in set theory. Not only is it unsolved; there are indications that it may even be *unsolvable* in a very strong sense.

Let us consider a hypothetical unsolved problem. Since this will be our substitute for a real problem, let us call it the *Virtual Problem*. Let us assume that the problem is to prove or refute the *Virtual Conjecture*. How would you like this:

Theorem 1: *The Virtual Problem is unsolvable.*

Well, if you have been working hard on the Virtual Problem without much luck, Theorem 1 may be a consolation prize. But could one possibly prove a theorem like Theorem 1? Yes and no. Intuitively speaking, if all of mathematics can be formalized in set theory, then also all modes of mathematical reasoning can be formalized. Thus, "the collected reasonings of all mathematicians of all times" become a mathematical object, and can be studied like any other mathematical object. It may be possible to prove that neither a proof nor a refutation of the Virtual Conjecture is among "the collected reasonings of all mathematicians of all times." Does this constitute a proof of Theorem 1? Almost. The assumption that all mathematics can be formalized in set theory is an act of belief that does not lend itself to mathematical scrutiny. While we can be reasonably sure that set theory encompasses essentially all correct mathematical arguments that have been used by mathematicians up to this point in history[1], there is always the somewhat remote possibility that eventually somebody will discover an immediately recognizable mathematical truth that transcends set theory. Thus, if it can be established that neither a proof nor a refutation of the Virtual Conjecture is among "the collected reasonings of all mathematicians of all times," something like the following theorem will have been proved:

Theorem 2: *The Virtual Conjecture is unsolvable, unless currently used foundations of mathematics are changed.*

Over the last three decades, set theorists have proved hundreds of theorems like Theorem 2. Such theorems are called *independence results.* The Virtual Conjecture does not have to be a strictly set-theoretical statement. It may be a problem in topology, algebra, functional analysis, or measure theory.

On the other hand, the foundations of mathematics have remained remarkably stable. Most mathematicians accept the axiomatic version ZFC[2] of set theory as a reasonably good foundation of mathematics and see little reason to exchange it for something else.[3] Thus, evidence is accumulating that many problems in set theory and related fields may be unsolvable in an absolute sense. However, if they are, we can never be entirely sure of this.

Set theory not only serves mathematics by providing a foundation and allowing one to delineate the limits of the knowable. It also *is* good mathematics. Set-theoretic theorems and techniques can be used in many other branches of mathematics much in the same way as linear algebra is used in differential equations. This is true not only for the concepts and methods known already to Cantor, but also for more recent results like Zorn's Lemma, the Erdős–Rado Theorem, or the Pressing Down Lemma. Volume II of this book is devoted to a presentation of set-theoretic techniques that are frequently applied outside of set theory. At the time

[1]Of course, this becomes immediately a false statement if the deliberately vague term "set theory" is replaced by a formal incarnation of it, like ZFC. In this case, the "immediately recognizable mathematical truth that transcends ZFC" could be the assertion that ZFC is consistent. However, if the "Virtual Conjecture" is something like the Continuum Hypothesis, then the intuitive picture drawn here will do for a reasonably accurate first approximation of the notion of an independence result.

[2]The letters stand for **Z**ermelo and **F**raenkel, who developed the system, and for one of the axioms, called the Axiom of **C**hoice.

[3]Not all mathematicians share this view. Mathematics can be developed in other frameworks. Some of these are brands of set theory similar to the version ZFC discussed in this text; others are entirely different approaches. In this book, we concentrate almost exclusively on a presentation of ZFC. (The only exceptions are occasional discussions of set theory without the Axiom of Choice.) We are far from claiming superiority of ZFC over alternative foundations of mathematics. For whatever reason, it won the competition. It does a decent job; so let us stick to it. It should be pointed out though that, to the best of our knowledge, none of the competitors of ZFC resolves the question of truth or falsity of CH, SH, MA, $\diamondsuit$, or of any other statement whose independence of ZFC has been established by the method of forcing.

of this writing, the use of elementary submodels seemingly is becoming popular with general topologists. In order to promote this trend, we decided to include a section on elementary submodels.

The history of set theory has not always been a smooth ride. In fact, the start was rather bumpy. While some mathematicians embraced set theory eagerly, others were openly hostile. The two quotes at the beginning of this Introduction are soundbites of rhetoric from the early days of set theory.

The power of the set concept lies in the possibility of treating collections of infinitely many objects m as a single entity M. Many contemporaries of Cantor felt uneasy about this approach. The question as to whether infinity actually exists, or is just an abstraction, a remote possibility that can be considered and approximated, but never attained, is as old as philosophy itself. For the Greek philosopher Plato, infinity was as real[4] as any finite object. His disciple Aristotle took the opposite stand: Infinity exists only as a potential that is never actually attained. The chasm between the Platonist and the Aristotelian approaches has permeated philosophical thought ever since. In essence, Cantor's treatment of infinity followed Plato, whereas his opponents espoused Aristotelian thinking. Nihil novi . . . , —and yet, for a brief moment in history the Platonist case seemed lost. In 1901, Bertrand Russell discovered that Cantor's definition of a set leads to a contradiction.

Let us say that an object x has property $\mathcal{P}$, if x is a set, but x is not an element of itself, which will be denoted by $x \notin x$. Let us collect all objects x with property $\mathcal{P}$ into a set M. Does M have property $\mathcal{P}$? Well, suppose M does not have property $\mathcal{P}$. Then it is not the case that $M \notin M$, i.e., $M \in M$. But, as an element of M, the set M must have property $\mathcal{P}$, which contradicts our assumption. If we assume that M has property $\mathcal{P}$, then M would have to be an element of M, i.e., M would have to be an element of itself, which contradicts the assumption that M has property $\mathcal{P}$. Thus, it is impossible for M to have property $\mathcal{P}$, and it is impossible for M to not have this property. This dilemma is called *Russell's Paradox.*

EXERCISE 2(PG) (a) Go carefully over the above argument and see whether our reasoning is perhaps wrong or incomplete.

(b) Should you find a gap in the reasoning, fill in the missing details, and see whether you still get a paradox.

Can one resolve Russell's Paradox, or do we have to accept it as a refutation of set theory? Of course it can be resolved; otherwise this book would not exist. Let us go back to Exercise 2(a). As you noticed, we forgot to check whether M is a set. Property $\mathcal{P}$ has two clauses. *If* M is a set, then M has property $\mathcal{P}$ iff $M \notin M$; but if M is not a set, then M does not have property $\mathcal{P}$, period. In the latter case, the paradox would disappear.

But is M a set? Let us consider Cantor's definition. In the spirit of his theory, collections of definite, distinct objects of our thought are sets. M satisfies the criterion of distinctness, since its members are distinguished from its nonmembers by a certain property. Thus, Cantor's definition implies that M is a set, and we get Russell's Paradox. (If you postponed Exercise 2(b), do it right now.)

We have seen that Russell's Paradox disappears if M is not a set. We have also seen that Cantor's definition implies that M is a set. Is there perhaps something wrong with Cantor's definition?

[4]In a sense, infinity was even more real for Plato than the finite objects of our perception.

To see its flaw, let us reexamine the process by which the set M of our example was constructed. M was the collection of certain definite objects of our thought, distinguished by a property $\mathcal{P}$ .... But wait a minute, how "definite" are these objects? If M might or might not be one of those, isn't M a bit indefinite? So, maybe, we should disqualify M as a possible element of M on the grounds of its indefiniteness?

EXERCISE 3(R): Try this and convince yourself that you get a similar paradox; this time with the added clause of some vaguely understood "definiteness" in the defining property of M.

But perhaps Cantor's definition could be salvaged by giving a precise meaning to the word "definite?" Think of the elements of a set as building blocks, and the formation of a set as assembling these building blocks into a whole. It is reasonable to require that at the moment a given set M is being formed, all its building blocks must have already attained their final shape; in this sense they should be "definite." Let us call this stance the *architect's view of set theory.* It stipulates that although it is possible to contemplate all sets at once, each set has to be formed at some moment in an abstract "time," and at that moment, all its building blocks must already have been available in their final shape.[5] Also, once a set is formed, one should be able to use it as a building block of other sets.

This view solves Russell's Paradox in an unexpected way: M is not a set, because it could never have been assembled! At no moment in set-theoretic time do all the building blocks for the construction of M exist.

How can the architect's view of set theory be expressed with sufficient mathematical precision? The approach concentrates not on what sets are, but on how sets are being formed. At the beginning of set-theoretic time, the only set that can be formed is the empty set, since no previous building blocks exist. Once this set is formed, it can be used as a building block for further sets. The modern alternative to Cantor's definition is to describe precisely by which operations new sets can be built from existing ones, and then to apply these operations successively to the empty set.

Can we get all sets in this way? Perhaps not, but we can construct a universe of sets rich enough to encompass all known mathematics. This will do for starters.

The architect's view of set theory can be formalized by axioms, similar to the way in which our space intuitions were formalized by Euclid more than two thousand years ago. The axiom system ZFC that will be studied in this book was proposed by E. Zermelo and A. Fraenkel early in this century. Once an axiom system has been formulated, one can ask whether a given mathematical statement or its negation follows from the axioms. The answer may be a "yes," a "no," or an independence result.

One often talks about "naive" versus "axiomatic" set theory. This may suggest a much deeper partition than there actually is.

EXERCISE 4(R): Does the Continuum Hypothesis[6] belong to naive or to axiomatic set theory?

---

[5]Note that this view is a synthesis of Platonist and Aristotelian elements. We shall see in Chapter 12 how the "timeline" of the set-theoretic universe looks.

[6]If you do not know what the Continuum Hypothesis is, ignore this exercise.

The relation between Cantor's view of sets and the axiomatic treatment of the theory is similar to that between our space intuitions and Euclidean geometry. The Euclidian axioms allow us to derive mathematical truths about points, lines and planes deductively. This is important, because the deductive method accounts for the high confidence mathematicians have in the truth of their theorems. However, our "naive" space intuitions are a valuable guide in guessing these theorems and outlines of their proofs. Also, similarities between geometric objects and features of the real world are grasped by our intuition, not by deductive reasoning. Without naivity, there would be no applications of mathematics.

When we say that somebody practices naive set theory, all we mean is that her arguments are based on Cantor's definition of a set. The naive approach is quite often the most enlightening one. If a mathematician's reasonings are also informed by a careful analysis of how sets are being built, we say that she practices axiomatic set theory. Frequently, this will just mean adopting the architect's point of view without being concerned about details of the axiomatization.[7]

In this book, both modes of set-theoretical thought will be practiced. We start out with a naive treatment of some of the basics: relations, functions, equipotency, order types and induction. As we go along, questions will arise that call for a more careful scrutiny. Chapters 5 and 6 contain a discussion of the axiomatic method, and in Chapter 7 we introduce the axioms. Next, we reexamine some of the material developed in the first four chapters from an axiomatic point of view. In Volume II, we discuss some set-theoretic results and techniques that we consider particularly important from the point of view of other mathematical disciplines. There we shall frequently switch from the naive to the axiomatic stance, and vice versa.

Of course, the choice of topics reflects our own biases and our desire to keep the number of pages finite.

Mathographical Remarks

The quote of Hilbert is taken from an address delivered on June 4, 1925. The text is printed in *Über das Unendliche*, Math. Ann. 95 (1926), 161–190.

The quote of Poincaré is taken from his article *L'Avenir des mathématiques*, Atti del IV Congresso Internazionale dei Matematici. Rome, 6–11 April 1908, Rome, Tipografia della R. Academia dei Lincei, C. V. Salviucci, 1909, 167–182. We thank David Fremlin for pointing out that Poincaré's views were misquoted in the first printing of this book.

Cantor's famous definition of a set is the first sentence of the article *Beiträge zur Begründung der transfiniten Mengenlehre*, Part I, Math. Ann. 46 (1895), 481–512.

Detailed accounts of the history of set theory in general, and of Cantor's work in particular can be found in the following books:

Joseph Warren Dauben, *Georg Cantor. His Mathematics and Philosophy of the Infinite*, Princeton University Press, 1990.

Michael Hallett, *Cantorian Set Theory and Limitation of Size*, Clarendon Press, Oxford, 1984.

Gregory H. Moore, *Zermelo's Axiom of Choice: Its Origins, Development and Influence*, Springer-Verlag, 1982.

[7]The use of the phrase "axiomatic set theory" in such instances may not be entirely appropriate. But it is commonly used, and there is no need to further complicate the picture by naming additional modes of practicing set theory.

English translations of many of the most influential papers on the foundations of mathematics written between 1879 and 1931 are reprinted in the book *From Frege to Gödel: A Source-book in Mathematical Logic,* 1879–1931, by Jean van Heijenoort, Harvard University Press, 1967.

The question to what extent Cantor's personal view of sets included elements of what we call "the architect's view" is a fascinating topic for philosophers and historians of science. If you are interested in this issue, we recommend the book *Understanding the Infinite,* by Shaughan Lavine, Harvard University Press, 1994.

Bertrand Russell found his Paradox in June 1901. He described it in a letter to Frege written June 16, 1902, and apparently also in an earlier letter to Peano. The paradox was first published in his book *The Principles of Mathematics,* vol. 1, Cambridge University Press, 1903.

The original version of the paradox was different from ours. We gave here a formulation of Russell's Paradox that very naturally leads to its resolution. If you want to imagine the impression Russell's Paradox must have made on his contemporaries, please keep in mind that they were lacking the benefit of hindsight which informed our choice of wording.

Part 1

Not Entirely Naive Set Theory

CHAPTER 1

Pairs, Relations, and Functions

EXERCISE 1(G): Consider the sets $a = \{\{\emptyset\}, \{\{\emptyset\}\}\}$, $b = \{\{\{\emptyset\}\}, \{\emptyset\}\}$, and $c = \{\{\{\emptyset\}\}, \{\emptyset\}, \{\{\emptyset\}\}\}$. Is any of these sets empty? Are these sets equal, or are they different? Is $\emptyset$ an element of a? Or perhaps a subset? How about $\{\emptyset\}$?

The empty set $\emptyset$ does not have any element; but this does not prevent it from being an element of another set, for instance of the set $\{\emptyset\}$. This set in turn can be an element of another set; in particular, it is an element of the sets a, b, c of Exercise 1. Thus, none of these sets is empty.

Are they different? Let us first consider another example. Suppose your instructor asked you to determine the union of the set of basketball players and the set of joggers in your class. Let us assume that John and Mary are joggers, and Mary is also the only basketball player in your class. You may answer the question by writing $\{John, Mary\}$. Your classmate Bill, who is very fond of Mary, will of course write $\{Mary, John\}$. A more methodical student may write down the set of joggers followed by the set of basketball players $\{Mary, John\}, \{Mary\}$, and then erase the middle pair of braces to get the answer $\{Mary, John, Mary\}$.

Is any of these answers more correct than the other two? Of course not. They are just different ways to denote the same set. The first two may be a little more elegant than the third, but all three are correct. It should be clear from this example that the order in which the elements of a set are given does not matter, nor does it matter if elements are repeated. The first property of sets that will be elevated to the status of an axiom (the so-called *Axiom of Extensionality*) is the following: *Two sets are equal iff they have the same elements.*

Thus, in Exercise 1, $a = b = c$.

EXERCISE 2(G): Why do we speak of *the* empty set rather than of *an* empty set?

The elements of a are $\{\emptyset\}$ and $\{\{\emptyset\}\}$. Since $\emptyset$ is not equal to any one of these two, it is not an element of a. However, because $\emptyset$ has no elements whatsoever, all elements of $\emptyset$ are by default also elements of a. As you know, this property is usually abbreviated by $\emptyset \subseteq a$. In this book, we shall use the symbol $\subset$ only if we want to emphasize that one set is a proper subset of another. We could thus write $\emptyset \subset a$, but it would be wrong to write $a \subset a$. On the other hand, $\{\emptyset\}$ is not a subset of a, since its only element $\emptyset$ is not a member of the latter.

We have seen that reversing the order of elements of a set does not cause identity crises for sets. But sometimes order does matter. Suppose your friends often help each other out in financial difficulties. To keep track of the borrowing that has been going on, you make a list. Instead of "Kathy owes Pam some money," you write shorthand $\langle Kathy, Pam\rangle$. It does matter whether you write $\langle Kathy, Pam\rangle$ or $\langle Pam, Kathy\rangle$, doesn't it?

Now we have the first challenge to our claim that all mathematics can be expressed in set-theoretical terms. Is it possible to have a set that would convey the same information as the ordered pair $\langle Kathy, Pam\rangle$? Of course, $\{Kathy, Pam\}$ is no good, because this is the same as $\{Pam, Kathy\}$. But there is a trick.

EXERCISE 3(G): Let a, b be any objects. Show that

$$\{\{a\},\{a,b\}\} = \{\{c\},\{c,d\}\} \quad \text{iff} \quad a = c \text{ and } b = d.$$

We define $\langle a, b\rangle = \{\{a\},\{a,b\}\}$, and call $\langle a, b\rangle$ an *ordered pair* as opposed to the *unordered pair* $\{a, b\}$. What we have done may look artificial. We could have defined $\langle a, b\rangle$ the other way round, or perhaps in an entirely different way. And besides, although Exercise 3 shows that our notion of ordered pair has the desired properties, can we really say that $\{\{a\},\{a,b\}\}$ *is* the ordered pair $\langle a, b\rangle$? Well, in what sense *is* $\langle Kathy, Pam\rangle$ the statement that "Kathy owes Pam some money?" Only to the extent that you chose it as a symbol for this statement, or rather, for the relationship concerning debt between Kathy and Pam. The symbol is appropriate, since it captures all the relevant information. In the same way, by defining $\langle a, b\rangle$ as $\{\{a\},\{a,b\}\}$, we make an agreement on how we shall communicate certain phenomena in the language of set theory. We are not making any ontological claims, i.e., we do not stipulate that $\langle a, b\rangle$ really *is* $\{\{a\},\{a,b\}\}$ in some deep philosophical sense.

Given three objects a, b, c, we define the ordered triple $\langle a, b, c\rangle$ by $\langle\langle a, b\rangle, c\rangle$. More generally, for $n \geq 2$, we define the ordered n-tuple of n objects $a_0, \ldots, a_{n-1}$ by $\langle \ldots \langle\langle a_0, a_1\rangle, \ldots\rangle, a_{n-1}\rangle$. The *Cartesian product* of n sets $A_0, \ldots, A_{n-1}$ is the set

$$A_0 \times \cdots \times A_{n-1} = \{\langle a_0, \ldots, a_{n-1}\rangle : a_i \in A_i \text{ for each } i < n\}.$$

EXERCISE 4(PG): (a) Let A, B be nonempty sets. Show that $\bigcup\bigcup(A \times B) = A \cup B$.

(b) Show that in (a) the assumption that A and B are nonempty can be slightly weakened, but not entirely omitted.

EXERCISE 5(PG): Show that two ordered triples $\langle a_0, a_1, a_2\rangle$ and $\langle b_0, b_1, b_2\rangle$ are equal iff $a_i = b_i$ for all $i < 3$.

Let us go back to the monetary relationships of your friends Kathy, Pam, John and Paul. Your list of who owes money to whom may look like this:

$$\langle Kathy, Pam\rangle, \langle Kathy, John\rangle, \langle Paul, Pam\rangle, \langle Paul, John\rangle.$$

If $F = \{Kathy, Pam, John, Paul\}$ is the set of your friends, then their "indebtedness relation" is a subset of $F \times F$. More generally, if A is a set, then any subset $R \subseteq A \times A$ will be called a *binary relation on A*. It does not matter whether or not R represents anything as interesting as indebtedness.

The indebtedness relation may not carry enough information for you. You may also want to know how much your friends owe each other. An appropriate list may look like this:

$$\langle Kathy, Pam, 125\rangle, \langle Kathy, John, 63\rangle, \langle Paul, Pam, 87\rangle, \langle Paul, John, 14\rangle.$$

The new relation is a subset of the Cartesian product $F \times F \times \mathbb{Z}$. Let us call such relations 3*–ary relations.* It is now easy to imagine n–ary relations for even larger n.

To a set theorist, a *function* $f : A \to B$ from a set A into a set B is simply a relation[1] $f \subseteq A \times B$ such that for each $a \in A$ there is exactly one $b \in B$ such that $\langle a, b\rangle \in f$. It is common to write $f(a) = b$ instead of $\langle a, b\rangle \in f$. The set A is called the *domain of* f and is denoted by $dom(f)$. Note that in particular, the empty set is a function with empty domain. For $C \subseteq A$ and $D \subseteq B$, we write $f[C] = \{b \in B : \exists a \in C\,(f(a) = b)\}$ for the *image of* C *under* f, and $f^{-1}D = \{a \in A : f(a) \in D\}$ for the *inverse image of* D. The image of the whole domain is called the *range of* f and is denoted by $rng(f)$. The function f is *onto* B or a *surjection,* if its range is equal to B. It is *one-to-one* or an *injection* if $f(a) \neq f(a')$ whenever $a \neq a'$. A function that is both a surjection and an injection is called a *bijection.*

Note that if f is a function from A into B, and if $B \subset C$, then f is also a function from A into C. But if $f : A \to B$ is a surjection, then $f : A \to C$ is *not* a surjection, although the f in both cases is the same set! As you can see, the word "surjection" is ambiguous. Therefore, we prefer to say that f maps A onto B, and use the word "surjection" only sporadically, when no confusion can arise.

One handy use of functions is the indexing of families of sets. Suppose I is a set, $\mathcal{A}$ is a family of sets, and $f : I \to \mathcal{A}$ is a surjection. If we denote $f(i) = A_i$, then $\mathcal{A} = \{A_i : i \in I\}$. Thus $\mathcal{A}$ becomes an *indexed family of sets.* Although the sets $\mathcal{A}$ and $\{A_i : i \in I\}$ are the same, most people find it more convenient to work with indexed families. For instance, the symbols $\bigcap \mathcal{A}$ and $\bigcap_{i\in I} A_i$ mean exactly the same thing, and yet the former notation sometimes causes confusion.

EXERCISE 6(G): Let $\mathcal{A}$ be a nonempty family of subsets of a set X, and let $\mathcal{B}$ be a nonempty family of subsets of a set Y. Moreover, let f be a function from X into Y. Show that

(a) $f^{-1}\bigcup\mathcal{B} = \bigcup\{f^{-1}B : B \in \mathcal{B}\}$;
(b) $f^{-1}\bigcap\mathcal{B} = \bigcap\{f^{-1}B : B \in \mathcal{B}\}$;
(c) $f^{-1}(Y\backslash B) = X\backslash f^{-1}B$ for each $B \in \mathcal{B}$.

EXERCISE 7(G): In the notation of the previous exercise, is it always true that

(a) $f[\bigcup\mathcal{A}] = \bigcup\{f[A] : A \in \mathcal{A}\}$;
(b) $f[\bigcap\mathcal{A}] = \bigcap\{f[A] : A \in \mathcal{A}\}$;
(c) $f[X\backslash A] = Y\backslash f[A]$ for each $A \in \mathcal{A}$?

Which inclusions always hold? What can be said if f is assumed to be onto? What if it is assumed one-to-one?

Here is another use of indexed families: The set of all functions from A into B is denoted by ${}^{A}B$. Given $\{A_i : i \in I\}$, we can define the *Cartesian product*

$$\prod_{i\in I} A_i = \{f \in {}^{I}(\bigcup\{A_i : i \in I\}) : \forall i \in I\,(f(i) \in A_i)\}.$$

[1]If we talk about a relation without mentioning its arity, we always mean a binary relation.

EXERCISE 8(PG): Let $A_\emptyset = \{\emptyset\}, A_{\{\emptyset\}} = \{\emptyset, \{\emptyset\}\}$. Using only braces and the empty set symbol, list all elements of the sets $A_\emptyset \times A_{\{\emptyset\}}$ and $\prod_{i\in\{\emptyset,\{\emptyset\}\}} A_i$.

As you can see, the two sets of Exercise 8 are not the same. However, both sets $A_\emptyset \times A_{\{\emptyset\}}$ and $\prod_{i\in\{\emptyset,\{\emptyset\}\}} A_i$ are called "Cartesian products." Although the two sets are formally different, there exists a canonical one-to-one map H from $\prod_{i\in\{\emptyset,\{\emptyset\}\}} A_i$ onto $A_\emptyset \times A_{\{\emptyset\}}$ which is defined by $H(f) = \langle f(\emptyset), f(\{\emptyset\})\rangle$. Therefore, we shall use the phrase "Cartesian product" in both of its meanings. In particular, the symbol A^2 may stand for $A \times A$ or for ${}^2A$. In the rare cases where it matters, we shall specify which of the two interpretations of this symbol is meant. For $n > 2$, unless specified otherwise, the symbol A^n refers to the set ${}^nA$, where $n = \{0, 1, \dots, n-1\}$.[2]

Let us go back to the set $F = \{Kathy, Pam, John, Paul\}$ of your friends. Let D denote the indebtedness relation on F that was considered above. This relation has the following property: If $\langle x, y\rangle \in D$, then $\langle y, x\rangle \notin D$. If a relation has this property, then it is called *asymmetric.* In contrast, a relation R on a set X is *antisymmetric* if for all $x, y \in X$, $\langle x, y\rangle \in R$ and $\langle y, x\rangle \in R$ implies that $x = y$. A relation R on a set X is *symmetric* if for all $x, y \in X$, whenever $\langle x, y\rangle \in R$ then also $\langle y, x\rangle \in R$. An example of a symmetric relation on the set F is the "same gender relation" S: a pair $\langle x, y\rangle \in S$ iff x and y are of the same gender. Thus

$$S = \{\langle Kathy, Pam\rangle, \langle Pam, Kathy\rangle, \langle John, Paul\rangle, \langle Paul, John\rangle,$$
$$\langle Kathy, Kathy\rangle, \langle Pam, Pam\rangle, \langle John, John\rangle, \langle Paul, Paul\rangle\}.$$

This relation has two more interesting properties:

- It is *reflexive,* i.e., $\langle x, x\rangle \in S$ for all $x \in F$.
- It is *transitive,* i.e., if both $\langle x, y\rangle \in S$ and $\langle y, z\rangle \in S$, then $\langle x, z\rangle \in S$.

A relation R on a set X is *irreflexive* if $\langle x, x\rangle \notin R$. Note that a nonreflexive relation is not necessarily irreflexive.

EXERCISE 9(G): Define a relation P on F by putting $\langle x, y\rangle \in P$ iff x and y are of opposite gender. Show that this relation is symmetric, but irreflexive and not transitive.

A binary relation that is reflexive, symmetric and transitive is called an *equivalence relation.* Here is an example of an equivalence relation on the set of all natural numbers $\mathbb{N}$:

$$\langle x, y\rangle \in C_6 \text{ iff there are } k, \ell, m \in \mathbb{N} \text{ such that } m < 6,$$
$$6k + m = x \text{ and } 6\ell + m = y.$$

In other words, $\langle x, y\rangle \in C_6$ if and only if these two numbers are congruent mod 6.

Given an equivalence relation R on a set X, we can form the set of *equivalence classes of R.* For $x \in X$, let the equivalence class of x be the set $x/_R = \{y \in X : \langle x, y\rangle \in R\}$. The element x is called a *representative of the equivalence class* $x/_R$. We also denote $X/_R = \{x/_R : x \in X\}$. Note that the equivalence classes of the relation C_6 are the congruence classes mod 6, and the equivalence classes of the "same gender relation" are the set of all females in the group and the set of all males in the group.

[2]This treatment of the natural numbers as sets will be discussed in Chapter 4.

EXERCISE 10(G): (a) Let R be an equivalence relation on a set X. Prove that $X/_R$ is a *partition of* X, i.e., that this family has the following two properties:

(1) $\bigcup(X/_R) = X$.

(2) For all $A, B \in X/_R$ either $A = B$ or $A \cap B = \emptyset$.

(b) Show that if $\mathcal{X}$ is a partition of a set X, then there exists exactly one equivalence relation R on X such that $\mathcal{X} = X/_R$.

One can define relations and functions on a set $X/_R$ of equivalence classes in terms of representatives. For example, on $\mathbb{N}/_{C_6}$, we can define a relation L as follows: $\langle x/_{C_6}, y/_{C_6}\rangle \in L$ iff there are $k, \ell, m, n \in \mathbb{N}$ such that $x = 6k+m$, $y = 6\ell+n$, and $m < n < 6$. Or, one could define a function $sq : \mathbb{N}/C_6 \to \mathbb{N}$ by: $sq(x/_{C_6}) = m^2$, where $x = 6k + m$ for some $k, m \in \mathbb{N}$ with $m < 6$.

Using this kind of definition, one should be on guard against a common fallacy. To illustrate it, suppose we would like to define a function φ on $\mathbb{N}/_{C_6}$ by $\varphi(x/_{C_6}) = [\frac{x}{2}]/_{C_6}$, where $[z]$ denotes the greatest integer not exceeding z. Then the function value of φ for $4/_{C_6}$ would be $2/_{C_6}$, and the function value of φ for $10/_{C_6}$ would have to be $5/_{C_6}$. But this is a contradiction, since 4 and 10 are in the same equivalence class of C_6, whereas 2 and 5 are not.

Whenever we define a relation or function on a set of equivalence classes in terms of representatives, we need to verify that the definition does not depend on the particular choice of representatives.

EXERCISE 11(G): Verify that the function sq and the relation L given above are well defined. That is, prove that if $\langle x, x'\rangle \in C_6$ and $\langle y, y'\rangle \in C_6$, then

(a) $\langle x/_{C_6}, y/_{C_6}\rangle \in L$ iff $\langle x'/_{C_6}, y'/_{C_6}\rangle \in L$;

(b) $sq(x/_{C_6}) = sq(x'/_{C_6})$.[3]

Mathographical Remarks

Our notation is fairly standard, but in some cases, alternative notations are used with about the same frequency. Many authors use (x, y) to denote the ordered pair $\langle x, y\rangle$. The image of a set A under a function f can be denoted by $f(A), f''A, \vec{f}(A), f^{\rightarrow}(A), \vec{f}A, f^{\rightarrow}A$; similarly for the inverse image, with arrows reversed. One often sees A^B instead of ${}^BA$. Equivalence classes of a relation R can be rendered as $[x], [x]_R, \|x\|, \|x\|_R$. Also, if R is a binary relation, it is usually more convenient to write xRy instead of $\langle x, y\rangle \in R$. We shall follow the latter convention in most of this book.

After reading this chapter, you should be able to determine whether you know enough mathematics to benefit from this book. If reading the chapter and doing the exercises felt comfortable, you are probably prepared to work through the material ahead. "Feeling comfortable" doesn't mean that you did not have to think at all. In fact, feeling comfortable with thinking about mathematical problems is perhaps the most important prerequisite for reading this book.

But, if the preceding material gave you headaches, you may want to work through a more elementary text on set theory before you return to this book. A detailed coverage of the material in this section can be found in many books. We recommend the following:

---

[3]This somewhat informal notation is used instead of the more cumbersome phrase "The value of sq does not depend on whether x or x' is used for its computation."

Paul R. Halmos, *Naive Set Theory,* van Nostrand, 1960, and Springer-Verlag, 1991.

Keith Devlin, *Sets, Functions, and Logic,* Chapman & Hall, 1992.

Douglas Smith, Maurice Eggen, and Richard St. Andre, *A Transition to Advanced Mathematics,* Brooks/Cole, 1990.

If you feel that you know all the necessary concepts, but that it would not hurt to refresh your knowledge by doing a few more exercises, we recommend:

Wiktor Marek and Janusz Onyszkiewicz, *Elements of Logic and Foundations of Mathematics in Problems,* Kluwer Academic Publishers or Polish Scientific Publishers, 1982.

The latter has all the exercises you may want to do, and even a few more than that.

Finally, if you think this book is at the right level for you, but you would rather read one of the competing books on the market, please feel free to do so. For the obvious reasons, we do not recommend any of them.

CHAPTER 2

Partial Order Relations

If a binary relation R on a set X is simultaneously reflexive, antisymmetric and transitive, it is called a *partial order relation on* X. A binary relation S on a set X that is simultaneously irreflexive and transitive is called a *strict partial order relation on* X. Whereas a partial order relation corresponds to the phrase "smaller than or equal to," a strict partial order relation is the mathematical equivalent of "strictly smaller than."

Consider the following example:

(0) Let $X = \{x, y, z\}$, and let $R = \{\langle x, x\rangle, \langle x, y\rangle, \langle x, z\rangle, \langle y, y\rangle, \langle z, z\rangle\}$. Then R is a partial order relation on X, and the strict partial order relation corresponding to R is $S = \{\langle x, y\rangle, \langle x, z\rangle\}$.

EXERCISE 1(G): (a) Show that every irreflexive, transitive binary relation is asymmetric.

(b) Show that if R is a partial order relation on X, then

$$S = R\backslash\{\langle x, x\rangle : x \in X\}$$

is a strict partial order relation. Similarly, show that if S is a strict partial order relation on X, then $R = S \cup \{\langle x, x\rangle : x \in X\}$ is a partial order relation.[1]

It follows from the above exercise that to every theorem about partial order relations, there corresponds a theorem about strict partial order relations, and vice versa. Like most authors, we shall concentrate on the theory of partial order relations. However, occasionally we shall use the strict partial order version of a theorem on partial order relations, usually without comment.

Partial order relations are usually denoted by symbols that are more suggestive than letters, like $\leq, \preceq, \trianglelefteq$. The corresponding strict partial order relations are then rendered by $<, \prec, \triangleleft$. If $\leq$ is a partial order relation on X, then the pair $\langle X, \leq\rangle$ is called a *partial order,* or *p.o.* Similarly, a *strict partial order* is a pair $\langle X, <\rangle$, where $<$ is a strict partial order relation on X. For any strict p.o. $\langle X, <\rangle$, the corresponding p.o. is denoted by $\langle X, \leq\rangle$.

EXERCISE 2(G): Let X be any set. By $\mathcal{P}(X)$ we denote the *power set of* X, i.e., the family of all subsets of X.

(a) Prove that $\langle \mathcal{P}(X), \subseteq\rangle$ is a p.o.

(b) Prove that if $\langle Y, \leq\rangle$ is a p.o. and $Z \subseteq Y$, then $\langle Z, \leq \cap (Z \times Z)\rangle$ is also a p.o.

(Note that both $\leq$ and $Z \times Z$ are subsets of $Y \times Y$, and that the intersection of these sets is a binary relation on Z. In the sequel, we shall be less formal and write $\langle Z, \leq\rangle$ instead of $\langle Z, \leq \cap (Z \times Z)\rangle$ in similar situations.)

[1] We call R the *partial order relation corresponding to* S.

(c) Conclude that if $\mathcal{A} \subseteq \mathcal{P}(X)$, then $\langle \mathcal{A}, \subseteq \rangle$ is a p.o.

Let us consider a few more examples.

(1) On the set of reals $\mathbb{R}$, we denote by $\leq$ the usual order relation.

(2) On the set $\mathbb{N} \setminus \{0\}$ of positive natural numbers, we define a relation $\leq_1$ by

$$a \leq_1 b \quad \text{iff} \quad a|b.$$

The symbol "|" stands for "a divides b," which means that there is a positive integer k such that $a \cdot k = b$.

(3) On the set of reals $\mathbb{R}$, we define a relation $\leq_2$ by

$$a \leq_2 b \quad \text{iff} \quad b - a \text{ is a nonnegative integer.}$$

(4) On the set of integers $\mathbb{Z}$, we introduce a relation $\leq_3$ by

$$a \leq_3 b \quad \text{iff} \quad 0 \leq a \leq b \;\text{ or }\; b \leq a < 0 \;\text{ or }\; a < 0 \leq b.$$

This relation looks as follows:

$$-1 \leq_3 -2 \leq_3 -3 \leq_3 ... \leq_3 0 \leq_3 1 \leq_3 2...$$

EXERCISE 3(G): Convince yourself that the relations in Examples (0)–(4) are partial order relations.

Two elements x and y are *comparable* by a partial order relation $\preceq$ if $x \preceq y$ or $y \preceq x$. A p.o. $\langle X, \preceq \rangle$ is called a *linear order,* or *l.o.,* if every two elements of X are comparable by $\preceq$.

EXERCISE 4(G): Check which of the p.o.'s in Examples (0)–(4) are l.o.'s.

Let $\langle A, \leq \rangle$ be a p.o., and let $a \in A$. We say that a is *minimal,* if the only $b \in A$ such that $b \leq a$ is a itself. a is *maximal,* if the only $b \in A$ such that $a \leq b$ is a itself. a is *minimum,* if $a \leq b$ for all $b \in A$, and a is *maximum,* if $b \leq a$ for all $b \in A$.

In Example (0), x is both a minimal and a minimum element; and y and z are maximal, but not maximum elements. It is easy to see that each minimum element in a p.o. is minimal, and each maximum element is maximal.

EXERCISE 5(G): Show that if a p.o. has a minimum (maximum) element, then this element is the only minimal (maximal) element.

In general, a p.o. may have more than one minimal (maximal) element, or none whatsoever. For example, consider $\langle \mathbb{N} \setminus \{0\}, \leq_1 \rangle$. This p.o. has a minimum element, namely 1, but no maximal element. If we delete this minimum element, then we obtain the p.o. $\langle \mathbb{N} \setminus \{0, 1\}, \leq_1 \rangle$ which has infinitely many minimal elements (the prime numbers), and thus, has no minimum element.

EXERCISE 6(G): (a) Let Fin be the family of finite subsets of $\mathbb{N}$. Find all minimal, minimum, maximal, and maximum elements of the p.o.'s $\langle Fin, \subseteq \rangle$ and $\langle Fin \setminus \{\emptyset\}, \subseteq \rangle$.

(b) Show that in a l.o., each minimal element is minimum, and each maximal element is maximum.

(c) Construct a p.o. that has exactly one minimal element, but no minimum.

We introduce now one of the most important concepts of set theory. We shall present it in full generality.

Definition 1: Let R be a binary relation on a set X. We say that R is *wellfounded,* if for every nonempty subset $Y \subseteq X$ there exists a $z \in Y$ such that $\langle y, z \rangle \notin R$ for all $y \in Y \backslash \{z\}$. A relation R is *strictly wellfounded* if it is wellfounded and irreflexive.

In particular, a partial order relation $\preceq$ on a set X is wellfounded if for every nonempty $Y \subseteq X$, the p.o. $\langle Y, \preceq \rangle$ has a minimal element. (In the sequel, we shall simply say in such situations that Y has a minimal element. Also, we shall say that the p.o. $\langle X, \preceq \rangle$ is wellfounded if the relation $\preceq$ is.)

Note that a p.o. $\langle X, \preceq \rangle$ is wellfounded if and only if the corresponding strict partial order $\langle X, \prec \rangle$ is strictly wellfounded. Note also that while a relation R on a nonempty set X cannot be both a strict partial order relation and a partial order relation, every strictly wellfounded relation is also a wellfounded relation.

EXERCISE 7(G): Let P be the set of all polynomials in one variable x with coefficients in $\mathbb{R}$. Define a binary relation D on P by: $\langle p_0, p_1 \rangle \in D$ iff p_0 is the derivative of p_1. Clearly, D is neither transitive, nor reflexive, nor irreflexive. Show that D is a wellfounded relation on P.

EXERCISE 8(G): Suppose $n > 0$ and $a_0, \ldots, a_n$ are pairwise different elements of a set X. Let $R \subseteq X \times X$ be such that

$$\{\langle a_0, a_1 \rangle, \langle a_1, a_2 \rangle, \ldots, \langle a_{n-1}, a_n \rangle, \langle a_n, a_0 \rangle\} \subseteq R.$$

Show that the relation R is not wellfounded.

The following theorem gives an extremely useful characterization of strictly wellfounded relations.

Theorem 2: *Let R be a binary relation on a set X. Then R is strictly wellfounded iff there is no sequence $\langle x_n : n \in \mathbb{N} \rangle$ of elements of X such that $\langle x_{n+1}, x_n \rangle \in R$ for all $n \in \mathbb{N}$.*

EXERCISE 9(G): Formulate an analogue of Theorem 2 for wellfounded relations.

EXERCISE 10(PG): Prove Theorem 2.

In Chapters 4 and 9, we shall scrutinize the reasoning involved in the proof of the "if" direction of Theorem 2 (see Exercises 4.13 and 9.45). For the time being, just use whatever reasoning looks sound. It is a good idea to keep a copy of your proof for later reference.

A l.o. $\langle X, \preceq \rangle$ such that $\preceq$ is wellfounded is called a *wellorder,* or *w.o.* A strict p.o. $\langle X, \prec \rangle$ corresponding to a w.o. $\langle X, \preceq \rangle$ is called a *strict w.o.* The prototypical w.o. is $\langle \mathbb{N}, \leq \rangle$.

EXERCISE 11(G): Which of the p.o.'s in Examples (0)–(4) are wellfounded? In particular, which of the l.o.'s are w.o.'s?

Among the most important contributions of set theory to the tool kit of mathematics is the use of abstract wellorders in arguments that traditionally involved only $\langle \mathbb{N}, \leq \rangle$. We give now a first illustration of this principle. You are probably already familiar with the technique of "disjointifying" families of sets that are indexed by natural numbers.

EXERCISE 12(G): Suppose $\mathcal{A} = \{A_n : n \in \mathbb{N}\}$ is a family of sets indexed by natural numbers. Show that there exists a family $\mathcal{B} = \{B_n : n \in \mathbb{N}\}$ such that $B_n \subseteq A_n$ for all $n \in \mathbb{N}$, the sets B_n are pairwise disjoint, and $\bigcup_{n\in\mathbb{N}} B_n = \bigcup_{n\in\mathbb{N}} A_n$.

What if the family $\mathcal{A}$ is indexed by some other set than $\mathbb{N}$? Let us say, $\mathcal{A} = \{A_i : i \in I\}$, where I is an arbitrary index set. Can we again disjointify the sets A_i as in Exercise 12? In other words, is there a family $\mathcal{B} = \{B_i : i \in I\}$ such that $B_i \subseteq A_i$ for all $i \in I$, the sets B_i are pairwise disjoint, and $\bigcup_{i\in I} B_i = \bigcup_{i\in I} A_i$?

The answer is "Yes" if there exists a wellorder relation on I. To see this, assume that $\langle I, \preceq \rangle$ is a w.o. For each $i \in I$ define: $B_i = A_i \backslash \bigcup_{j\prec i} B_j$.

EXERCISE 13(G): Show that the sets B_i are pairwise disjoint, and that $\bigcup_{i\in I} B_i = \bigcup_{i\in I} A_i$.

So, if every set admits a wellorder, then every indexed family of sets can be disjointified. But does every set admit a wellorder? As we shall see in Chapter 9, this is indeed true in ZFC. It is far from being obvious though. Try to do the following exercise, but even if you are not easily discouraged, do not spend more than ten minutes on it.

EXERCISE 14(X): Without using the techniques of Chapter 9, find a wellorder relation on the set of reals $\mathbb{R}$.

Now let us return to Exercise 11. As you noticed, the p.o. $\langle \mathbb{N}\backslash\{0\}, \leq_1 \rangle$ of example (2) is wellfounded. What could a formal proof of this fact look like? Let us consider two types of arguments that will be useful in other situations as well.

Extension to w.o. argument: If $n \leq_1 m$, then $n \leq m$. In other words, the relation $\leq_1$ is a subset of the relation $\leq$. The latter is known to be a wellorder relation on $\mathbb{N}\backslash\{0\}$. Now wellfoundedness of $\leq_1$ is a consequence of the following simple fact.

Claim 3: *Suppose R, S are two binary relations on a set X such that $R \subseteq S$. If S is wellfounded, then so is R.*

EXERCISE 15(G): (a) Prove Claim 3. You may use Theorem 2.
(b) Conclude that if $\langle X, \preceq \rangle$ is a w.o., and if $Y \subseteq X$, then $\langle Y, \preceq \rangle$ is also a w.o.

You may wonder whether it is rather common for wellfounded relations to be contained in a wellorder relation, or whether it was sheer luck that the relation $\leq_1$

had this property. We shall investigate this matter in Chapter 9.

Rank function argument: This type of argument makes use of the following fact.

Claim 4: *Let R be a binary relation on a set X, and suppose $\langle Y, \prec \rangle$ is a strict w.o. If there exists a function $rk : X \to Y$ such that*

$$(*) \qquad \forall x, y \in X \, (x \neq y \land \langle x, y \rangle \in R \to rk(x) \prec rk(y)),$$

then the relation R is wellfounded.

In Chapter 4, a function that satisfies $(*)$ and always takes on the smallest possible value consistent with this requirement will be called a *rank function for* R. In the same chapter, we shall show that every wellfounded relation has a rank function; so the converse of Claim 4 is also true.

Now let us consider two examples of rank functions. For the p.o. $\langle X, R \rangle$ of Example (0), let $rk_0 : X \to \mathbb{N}$ be defined by: $rk_0(x) = 0$ and $rk_0(y) = rk_0(z) = 1$.

EXERCISE 16(G): Convince yourself that the function rk_0 satisfies condition $(*)$ of Claim 4.

Now consider the p.o. $\langle \mathbb{N}, \leq_1 \rangle$, and let $n \in \mathbb{N} \backslash \{0\}$. Then n has a unique representation $n = \prod_{k \in \mathbb{N} \backslash \{0\}} p_k^{i_k(n)}$, where p_k denotes the k-th prime, $i_k(n) \in \mathbb{N}$, and $i_k(n) = 0$ for all but finitely many k. Define: $rk_1(n) = \sum_{k \in \mathbb{N} \backslash \{0\}} i_k(n)$.

EXERCISE 17(G): Show that the function $rk_1 : \mathbb{N} \backslash \{0\} \to \mathbb{N}$ satisfies condition $(*)$ of Claim 4,

EXERCISE 18(G): Prove Claim 4.

Definition 5: Let $\langle X, \preceq_X \rangle, \langle Y, \preceq_Y \rangle$ be p.o.'s. A function $\pi : X \to Y$ is called *order-preserving,*[2] if $\forall x, y \in X \, (x \preceq_X y \leftrightarrow \pi(x) \preceq_Y \pi(y))$.

If, moreover, π maps X onto Y, then we say that π is an *order isomorphism.* If there exists an order-preserving function from X into Y, then we say that the p.o. $\langle X, \preceq_X \rangle$ can be *embedded in* $\langle Y, \preceq_Y \rangle$.

If there exists an order isomorphism between X and Y, then we say that the p.o.'s $\langle X, \preceq_X \rangle$ and $\langle Y, \preceq_Y \rangle$ are *isomorphic,* and write $\langle X, \preceq_X \rangle \cong \langle Y, \preceq_Y \rangle$.

For example, the function *arctan* is an embedding of $\langle \mathbb{R}, \leq \rangle$ into itself. It is also an order isomorphism between $\langle \mathbb{R}, \leq \rangle$ and $\langle (-\pi/2, \pi/2), \leq \rangle$.

If $X \subseteq Y$ and $\preceq_X = \preceq_Y \cap (X \times X)$, then the identity function $i : X \to Y$ that maps each element of X to itself is an order embedding. Of particular importance are situations where $x \prec_Y y$ for all $x \in X$ and $y \in Y \backslash X$. In this case we say that $\langle Y, \preceq_Y \rangle$ is an *end-extension* of $\langle X, \preceq_X \rangle$. For example, $\langle [0, 1], \leq \rangle$ is an end-extension of $\langle [0, 1), \leq \rangle$, but not of $\langle (0, 1), \leq \rangle$.

[2]Some authors require only that order-preserving maps satisfy the implication $x \preceq_X y \to \pi(x) \preceq_Y \pi(y)$.

EXERCISE 19(G): Show that every order-preserving function is one-to-one. Conclude that every order-preserving surjection is an order isomorphism.

We shall also speak of order-preserving maps, order embeddings, and order-isomorphisms between strict p.o.'s. In particular, if $\langle X, \prec_X \rangle, \langle Y, \prec_Y \rangle$ are strict p.o.'s, then a one-to-one function $\pi : X \to Y$ is called *order-preserving* if $\forall x, y \in X\, (x \prec_X y \leftrightarrow \pi(x) \prec_Y \pi(y))$.

EXERCISE 20(G): (a) Suppose that $\langle X, \preceq_X \rangle, \langle Y, \preceq_Y \rangle$ are p.o.'s and that $\pi : X \to Y$ is order-preserving in the sense of Definition 5. Show that π is also order-preserving with respect to the corresponding strict p.o.'s $\prec_X$ and $\prec_Y$.

(b) Give an example of strict p.o.'s $\langle X, \prec_X \rangle$ and $\langle Y, \prec_Y \rangle$ and of a function $\pi : X \to Y$ that is not one-to-one, and yet has the property that $\forall x, y \in X\, (x \prec_X y \leftrightarrow \pi(x) \prec_Y \pi(y))$.

(c) Show that if $\langle X, \prec_X \rangle$ and $\langle Y, \prec_Y \rangle$ are strict l.o.'s and $\pi : X \to Y$ is a function such that $\forall x, y \in X\, (x \prec_X y \leftrightarrow \pi(x) \prec_Y \pi(y))$, then π is one-to-one.

EXERCISE 21(PG): (a) Show that the p.o. $\langle Fin, \subseteq \rangle$ of Exercise 6 can be embedded in $\langle \mathbb{N} \backslash \{0\}, \leq_1 \rangle$.

(b) Show that for every p.o. $\langle X, \preceq \rangle$ there exists a family of sets $\mathcal{A}$ such that $\langle X, \preceq \rangle \cong \langle \mathcal{A}, \subseteq \rangle$. *Hint*: Let $\mathcal{A} \subseteq \mathcal{P}(X)$ be the family of sets $\mathcal{A} = \{\{y : y \preceq x\} : x \in X\}$.

The theorem proved in Exercise 21(b) is an example of a *representation theorem.* The theorem says that every (abstract) partial order relation can be represented by the (concrete) relation $\subseteq$ on a suitably chosen family of sets. Since it is easier to think about concrete examples than about abstract properties, representation theorems can be a powerful aid to our intuition.

If a p.o. $\langle X, \preceq_X \rangle$ can be embedded in a p.o. $\langle Y, \preceq_Y \rangle$, and $\langle Y, \preceq_Y \rangle$ can also be embedded in $\langle X, \preceq_X \rangle$, it still does not need to be true that $\langle X, \preceq_X \rangle$ and $\langle Y, \preceq_Y \rangle$ are isomorphic; even if both p.o.'s are linear. As an example, consider $\langle \mathbb{R}, \leq \rangle$ and $\langle (-\pi/2, \pi/2], \leq \rangle$. As we have seen above, the first of these l.o.'s can be embedded in the second; and the identity function is obviously an embedding of the second l.o. in the first. Nevertheless, the two l.o.'s are not isomorphic.

EXERCISE 22(G): Why?

A property that can be expressed entirely in terms of the partial order relation is called an *order property.* Order properties are *invariant under order isomorphisms* or *order invariants;* which means that if a p.o. $\langle X, \preceq_X \rangle$ has this property, and if $\langle X, \preceq_X \rangle \cong \langle Y, \preceq_Y \rangle$, then the p.o. $\langle Y, \preceq_Y \rangle$ also has this property. As the order isomorphism between $\langle \mathbb{R}, \leq \rangle$ and $\langle (-\pi/2, \pi/2), \leq \rangle$ shows, the property of being an interval of $\mathbb{R}$ of finite length is not an order property.

An example of an order property is the number of minimal or maximal elements. Since this number is zero for $\langle \mathbb{R}, \leq \rangle$ and one for $\langle (-\pi/2, \pi/2], \leq \rangle$, the two l.o.'s cannot be isomorphic. Another example of an order invariant is the property of being dense. A p.o. $\langle X, \preceq_X \rangle$ is *dense* if for all $x, y \in X$, if $x \prec_X y$, then there exists a $z \in X$ such that $x \prec_X z \prec_X y$. The l.o. $\langle \mathbb{Q}, \leq \rangle$ is dense, whereas the l.o. $\langle \mathbb{Q} \cap ((0, 1] \cup [2, 3)), \leq \rangle$ is not. It follows that these two l.o.'s are not isomorphic.

However, if we delete just one element from the last example, we get a l.o. that is isomorphic to $\langle \mathbb{Q}, \leq \rangle$.

EXERCISE 23(G): Show that $\langle \mathbb{Q}, \leq \rangle \cong \langle \mathbb{Q} \cap ((0,1) \cup [2,3)), \leq \rangle$.

At this point of the course, the best way to solve Exercise 23 is to construct an order isomorphism. In Chapter 4 we shall present a powerful theorem that allows one to solve Exercise 23 without explicitly constructing such a function.

EXERCISE 24(G): Let $\langle X, \preceq_X \rangle, \langle Y, \preceq_Y \rangle$ and $\langle Z, \preceq_Z \rangle$ be p.o.'s. Show that
(a) $\langle X, \preceq_X \rangle \cong \langle X, \preceq_X \rangle$;
(b) $\langle X, \preceq_X \rangle \cong \langle Y, \preceq_Y \rangle$ implies $\langle Y, \preceq_Y \rangle \cong \langle X, \preceq_X \rangle$;
(c) If $\langle X, \preceq_X \rangle \cong \langle Y, \preceq_Y \rangle$ and $\langle Y, \preceq_Y \rangle \cong \langle Z, \preceq_Z \rangle$, then $\langle X, \preceq_X \rangle \cong \langle Z, \preceq_Z \rangle$.

Thus, order isomorphism behaves like an equivalence relation. But what could be the set **PO** such that $\cong \subseteq \mathbf{PO} \times \mathbf{PO}$? The set of all p.o.'s of course. Oops, this is not a set.

EXERCISE 25(R): (a) Argue on the grounds of the architect's view of set theory (which was outlined in the Introduction) that **PO** cannot be a set.
(b) Moreover, show that for each given p.o. $\langle X, \preceq_X \rangle$, if $X \neq \emptyset$, then the collection of all p.o.'s $\langle Y, \preceq_Y \rangle$ that are order-isomorphic to $\langle X, \preceq_X \rangle$ is not a set.

Too bad. However, **PO** is of course some collection of sets. We shall see in Chapter 8 how collections like **PO** can be handled in axiomatic set theory. For now, let us just remark that collections like **PO** cannot be elements of sets. Other than that, we can talk about them in much the same way as we talk about sets. Whenever we are not sure whether a given collection of sets is a set, we call it a *class*. If we are sure that it is not a set, then we call it a *proper class*. As long as we keep in mind that it is a proper class, we can treat $\cong$ like an equivalence relation on **PO**. In order to indicate that $\cong$ is too big a collection to form a set, we call it an *equivalence* (instead of an equivalence relation).

If $\langle X, \preceq_X \rangle \cong \langle Y, \preceq_Y \rangle$, then we say that $\langle X, \preceq_X \rangle$ and $\langle Y, \preceq_Y \rangle$ *are of the same order type*. Order types of wellorders are called *ordinals*.

What kind of objects are "order types?" Notions of types abound in the sciences, and even in everyday life. "Nationality," "gender," and "age" are examples of types that can be used to classify humans. Most of the time we use these concepts without much reflection on what they represent. But, as you certainly noticed, mathematicians tend to be picky in such matters. In the previous section, we mentioned that "gender" can be understood as an equivalence class of a certain equivalence relation. Please convince yourself that "nationality" and "age" can also be understood in the same way, although the boundaries between the equivalence classes have to be either somewhat artificial, or a little fuzzy.[3] Thus, by analogy, we would like to define for each p.o. $\langle X, \preceq \rangle$ its order type $ot(\langle X, \preceq \rangle)$ (or just $ot(X)$) as the class of all p.o.'s isomorphic to it. As we have seen in Exercise 25(b), unless X is empty, this is a proper class.

[3] There is a version of set theory called "fuzzy set theory" that deals with situations where the membership relation is not crisply determined. This theory has many applications in computer science and robotics, but is beyond the scope of this text.

For now, thinking of $ot(X)$ as an equivalence class may give you the right intuitions. An alternative treatment of order types in general, and of ordinals in particular, will be discussed in Chapters 8, 10, and 12.

Order types are commonly denoted by lower case Greek letters. In particular, $ot(\langle \mathbb{N}, \leq \rangle) = \omega$, $ot(\langle \mathbb{Q}, \leq \rangle) = \eta$, and $ot(\langle \mathbb{R}, \leq \rangle) = \lambda$. In this terminology, Exercise 19 could be phrased as: "Show that $ot(\langle \mathbb{Q} \cap ((0,1) \cup [2,3)), \leq \rangle) = \eta$."

We show in Chapter 4 that the order types η and λ can be characterized as unique order types with certain properties.

Now we shall present some constructions of new p.o.'s from given ones.

Let $\langle A, \preceq_A \rangle$ and $\langle B, \preceq_B \rangle$ be p.o.'s. The *order sum* $\langle A, \preceq_A \rangle \oplus \langle B, \preceq_B \rangle$ (or simply $A \oplus B$ if the relations are implied by context) is the p.o. $\langle C, \preceq \rangle$, where

$$C = A \times \{0\} \cup B \times \{1\},$$

and the relation $\preceq$ is given by

$$\langle a, b \rangle \preceq \langle c, d \rangle \text{ iff } \begin{cases} b = 0 \text{ and } d = 1 \text{ or} \\ b = d = 0 \text{ and } a \preceq_A c \text{ or} \\ b = d = 1 \text{ and } a \preceq_B c. \end{cases}$$

Note that $A \times \{0\} \cap B \times \{1\} = \emptyset$. Thus one can think of $A \times \{0\}$ and $B \times \{1\}$ as disjoint copies of A and B.

EXERCISE 26(G): (a) Show that if $\langle A, \preceq_A \rangle$ and $\langle B, \preceq_B \rangle$ are wellfounded p.o.'s, then so is $\langle A, \preceq_A \rangle \oplus \langle B, \preceq_B \rangle$.

(b) Show that if $\langle A, \preceq_A \rangle$ and $\langle B, \preceq_B \rangle$ are l.o.'s, then so is $\langle A, \preceq_A \rangle \oplus \langle B, \preceq_B \rangle$.

(c) Suppose $\langle A, \preceq_A \rangle \cong \langle A', \preceq_{A'} \rangle$ and $\langle B, \preceq_B \rangle \cong \langle B', \preceq_{B'} \rangle$. Prove that

$$\langle A, \preceq_A \rangle \oplus \langle B, \preceq_B \rangle \cong \langle A', \preceq_{A'} \rangle \oplus \langle B', \preceq_{B'} \rangle.$$

The operation $\oplus$ induces an operation on order types. If $\alpha = ot(\langle A, \preceq_A \rangle)$ and $\beta = ot(\langle B, \preceq_B \rangle)$, then we define $\alpha + \beta = ot(\langle A, \preceq_A \rangle \oplus \langle B, \preceq_B \rangle)$. It follows from Exercise 26(c) that the operation $+$ is well defined. As an illustration, look at Example (4) from the beginning of this section. It is not hard to see that $ot(\langle \mathbb{Z}, \leq_3 \rangle) = \omega + \omega$.

Let $\langle A, \preceq \rangle$ be a p.o. We define a relation $\preceq^*$ on A by letting $a \preceq^* b$ iff $b \preceq a$.

EXERCISE 27(G): (a) Show that if $\langle A, \preceq \rangle$ is a p.o., then so is $\langle A, \preceq^* \rangle$.

(b) Show that if $\langle A, \preceq \rangle$ is a l.o., then so is $\langle A, \preceq^* \rangle$.

(c) Give an example of a w.o. $\langle A, \preceq \rangle$ such that $\langle A, \preceq^* \rangle$ is not a w.o.

(d) Show that if $ot(\langle A, \preceq_A \rangle) = ot(\langle B, \preceq_B \rangle)$, then $ot(\langle A, \preceq_A^* \rangle) = ot(\langle B, \preceq_B^* \rangle)$.

If $ot(\langle A, \preceq \rangle) = \varrho$, then ϱ^* denotes $ot(\langle A, \preceq^* \rangle)$. By Exercise 27(d), ϱ^* is well defined.

EXERCISE 28(G): (a) Show that $ot(\langle \mathbb{Z}, \leq \rangle) = \omega^* + \omega$.

(b) Show that if $ot(\langle A, \preceq \rangle) = \omega + \omega^*$, then $\langle A, \preceq \rangle$ has a minimum element.

(c) Conclude that addition of infinite order types is not a commutative operation.

One can meaningfully define three kinds of products of p.o.'s. The *simple product* $\langle A, \preceq_A\rangle \otimes^s \langle B, \preceq_B\rangle$ (or shorthand $A \otimes^s B$) is the p.o. $\langle A \times B, \preceq_s\rangle$, where the relation $\preceq_s$ is given by

$$\langle a, b\rangle \preceq_s \langle c, d\rangle \text{ iff } a \preceq_A c \text{ and } b \preceq_B d.$$

EXERCISE 29(G): (a) Verify that if $\langle A, \preceq_A\rangle$ and $\langle B, \preceq_B\rangle$ are p.o.'s, then so is $\langle A, \preceq_A\rangle \otimes^s \langle B, \preceq_B\rangle$.

(b) Give examples of l.o.'s $\langle A, \preceq_A\rangle$ and $\langle B, \preceq_B\rangle$ such that $\langle A, \preceq_A\rangle \otimes^s \langle B, \preceq_B\rangle$ is not a linear order.

The *lexicographic product* or *lexicographic order* $\langle A, \preceq_A\rangle \otimes^l \langle B, \preceq_B\rangle$ (or shorthand $A \otimes^l B$) is the p.o. $\langle A \times B, \preceq_l\rangle$, where the relation $\preceq_l$ is given by

$$\langle a, b\rangle \preceq_l \langle c, d\rangle \text{ iff } \begin{cases} a \prec_A c & \text{or} \\ a = c \text{ and } b \preceq_B d. \end{cases}$$

EXERCISE 30(G): Guess why this is called the lexicographic order.

The *antilexicographic product* or *antilexicographic order* $\langle A, \preceq_A\rangle \otimes^a \langle B, \preceq_B\rangle$ (or shorthand $A \otimes^a B$) is the p.o. $\langle A \times B, \preceq_a\rangle$, where the relation $\preceq_a$ is given by

$$\langle a, b\rangle \preceq_a \langle c, d\rangle \text{ iff } \begin{cases} b \prec_B d & \text{or} \\ b = d \text{ and } a \preceq_A c. \end{cases}$$

EXERCISE 31(G): Consider $\langle [0,1), \leq\rangle \otimes^s \langle [0,1], \leq\rangle$, $\langle [0,1), \leq\rangle \otimes^l \langle [0,1], \leq\rangle$, and $\langle [0,1), \leq\rangle \otimes^a \langle [0,1], \leq\rangle$. Find all minimal, minimum, maximal and maximum elements of these p.o.'s.

EXERCISE 32(PG): Prove the following:
(a) If $\langle A, \preceq_A\rangle$ and $\langle B, \preceq_B\rangle$ are l.o.'s, then so is $\langle A, \preceq_A\rangle \otimes^a \langle B, \preceq_B\rangle$.
(b) If $\langle A, \preceq_A\rangle$ and $\langle B, \preceq_B\rangle$ are w.o.'s, then so is $\langle A, \preceq_A\rangle \otimes^a \langle B, \preceq_B\rangle$.

The antilexicographic product is used to define a notion of multiplication of order types. If $\alpha = ot(\langle A, \preceq_A\rangle)$ and $\beta = ot(\langle B, \preceq_B\rangle)$, then we define

$$\alpha \cdot \beta = ot(\langle A, \preceq_A\rangle \otimes^a \langle B, \preceq_B\rangle).$$

EXERCISE 33(PG): Show that $\alpha \cdot \beta$ is well defined.

EXERCISE 34(PG): Prove that the operations on order types satisfy the following associativity and distributivity laws.
(a) $(\alpha + \beta) + \gamma = \alpha + (\beta + \gamma)$;
(b) $(\alpha \cdot \beta) \cdot \gamma = \alpha \cdot (\beta \cdot \gamma)$;
(c) $\alpha \cdot (\beta + \gamma) = (\alpha \cdot \beta) + (\alpha \cdot \gamma)$.

One can define order products involving infinitely many factors. We shall do this for the lexicographic product.

Let $\langle I, \preceq\rangle$ be a w.o., and for each $i \in I$ let $\langle A_i, \preceq_i\rangle$ be a p.o. Then $\bigotimes_{i\in I}^{l}\langle A_i, \preceq_i\rangle$ (or shorthand $\bigotimes_{i\in I}^{l} A_i$) is the p.o. $\langle D, \preceq_D\rangle$, where $D = \prod_{i\in I} A_i$ and $u \preceq_D v$ iff $u = v$ or $u(i) <_i v(i)$ for the $\preceq$-smallest $i \in I$ such that $u(i) \neq v(i)$.

EXERCISE 35(G): (a) Show that the relation $\preceq_D$ defined above is a partial order relation.

(b) Show that if each $\langle A_i, \preceq_i\rangle$ is a l.o., then so is $\langle D, \preceq_D\rangle$.

It is not in general true that an infinite product of w.o.'s is a w.o. To see this, let $\langle I, \preceq\rangle = \langle A_i, \preceq_i\rangle = \langle \mathbb{N}, \leq\rangle$ for all $i \in \mathbb{N}$. For $n \in \mathbb{N}$, let $u_n \in \prod_{i\in\mathbb{N}} \mathbb{N}$ be the function defined by:

$$u_n(i) = \begin{cases} 0 & \text{if } i \leq n, \\ 1 & \text{if } i > n. \end{cases}$$

EXERCISE 36(G): Show that the u_n's form an infinite decreasing sequence in $\bigotimes_{i\in\mathbb{N}}^{l} \mathbb{N}$. Conclude that this lexicographic product is not a w.o.

EXERCISE 37(PG): Define infinite versions of the simple product and the antilexicographic product. Pay special attention to the kind of order relation you need on the index set (if, in fact, you need any such relation.)

We conclude this chapter with a notion closely related to that of a partial order relation.

Definition 6: A relation R on a set A is called a *partial pre-order relation* if it is reflexive and transitive. If R is a partial pre-order relation on A, then the pair $\langle A, R\rangle$ is called a *partial pre-order.*

Every partial order relation is a partial pre-order relation, but not vice versa. For example, consider the relation $\subseteq^*$ on $\mathcal{P}(\mathbb{N})$ defined by:

$$a \subseteq^* b \text{ iff } a\backslash b \text{ is finite.}$$

EXERCISE 38(G): Show that $\subseteq^*$ is a partial pre-order relation on $\mathcal{P}(\mathbb{N})$, but that it is not a partial order relation.

With any partial pre-order relation R we associate another binary relation $\sim_R$ defined by:

$$a \sim_R b \text{ iff } aRb \text{ and } bRa.$$

EXERCISE 39(G): (a) Show that if R is a partial pre-order relation on a set A, then $\sim_R$ is an equivalence relation on A.

(b) Show that if $a \subseteq \mathbb{N}$, then $a/_{\sim_{\subseteq^*}} = \{b \subseteq \mathbb{N} : a\Delta b \text{ is finite}\}$.

For any given partial pre-order relation R on a set A we can define a relation $\leq_R$ on $A/_{\sim_R}$ by:

$$a/_{\sim_R} \leq_R b/_{\sim_R} \text{ iff } aRb$$

EXERCISE 40(G): Let R be any partial pre-order on a set A.

(a) Show that the relation $\leq_R$ is well defined.

(b) Show that $\langle A/_{\sim_R}, \leq_R \rangle$ is a p.o.

It is customary to denote the relation $\sim_{\subseteq^*}$ by $=^*$ and to make no notational distinction between R and $\leq_R$. In particular, the symbol $\subseteq^*$ is also used to denote $\leq_{\subseteq^*}$. The resulting ambiguity is more than offset by the reduction in the number of symbols. We shall follow this practice in Volume II, where properties of the p.o. $\langle \mathcal{P}(\mathbb{N})/_{=^*}, \subseteq^* \rangle$ will be studied.

EXERCISE 41(G): Let $\langle X, \preceq \rangle$ be a l.o., and let $f : X \to Y$. Define a relation R on Y as follows: $R = \{\langle f(x), f(x') \rangle : x, x' \in X \wedge x \preceq x'\}$.

(a) Show that R is transitive.

(b) Show that R is a partial pre-order relation on Y iff f maps X onto Y.

(c) Show that $\langle Y, R \rangle$ is a l.o. iff f is a bijection between X and Y.

(d) Show that if f is a bijection, then f is an order isomorphism between $\langle X, \preceq \rangle$ and $\langle Y, R \rangle$.

(e) Show that if $\langle X, R \rangle$ is any partial pre-order, then there exist a p.o. $\langle Y, \preceq \rangle$ and a map $f : X \to Y$ such that $R = \{\langle x, x' \rangle : x, x' \in X \wedge f(x) \preceq f(x')\}$.

Mathographical Remarks

Most authors do not carefully distinguish between "partial order" and "partial order relation." This rarely leads to confusion, but we felt that a beginner might be more comfortable with the terminology adopted here. You should be aware that in other texts, the phrase "partial order" may mean "partial order relation." Also, many authors use the name "partial order relation" for what we call "strict partial order relation." Similarly, the expression "wellfounded relation" is sometimes used exclusively for strictly wellfounded relations. This variety of terminological conventions may be confusing. But if you constantly keep in mind that there is nothing sacred about the words mathematicians use, it is not so difficult to figure out what a particular author has in mind.

Our notation for the different kinds of products is somewhat idiosyncratic; apparently, there is no commonly accepted notation. The use of the symbols η and λ for the order types of the rationals and the real line dates back to the classical writings of Cantor and Hausdorff. Recently, set theorists have not been studying these order types a lot, and the letter λ is much more often used to denote an infinite cardinal. Also, if you see an η, it probably stands for an ordinal. But, if the order types of the reals and of the rationals are being discussed, then chances are the notation of this section will be used.

The use of partial orders instead of linear orders has recently been on the increase in set theory. A presentation of this trend will be an underlying theme of this book. That is why we introduced the general notion of wellfoundedness, rather than just restricting ourselves to wellorders, as most introductory texts do.

Finally, a word on hyphenation. Mathematicians are divided on whether to write wellorder, well-order, or well order. We decided in favor of wellorder, which looks reasonably correct to us.

CHAPTER 3

Cardinality

What is "three"? This question looks childishly simple; the answer "A natural number" comes immediately to mind. All right, but what *is* a natural number? Hm. We know that you *know* what "three" means. Perhaps, you just can't offhand find a definition suitable for a graduate level mathematics class. This is not surprising, since you learned the concept in kindergarten. Let us forget for a moment about graduate level. "Three" is the property that three kids, three apples, and three teddybears have in common. Believe it or not, the leap of imagination required to pass from "three teddybears" to plain "three" is much bigger than anything we are going to demand of you in this book.

Let us see whether we can express your subconscious mathematical insight in a formal definition. What makes three apples different from four apples, or from two apples? Suppose there are three kids to each of whom you want to give an apple out of a bag. You start passing out the apples, one at a time, and different apples to different kids. If each kid got an apple, and you are left with an empty bag, then you must have had three apples in the bag to begin with. If each kid got an apple, and the bag is not empty, there must have been more than three apples in the bag. No problem here; you can have one yourself. If you start with fewer than three apples, then some kid will get no apple at all. This is disaster.

You may object that we are trying to define "three" in terms of "three." You are wrong: We only want to explain "three apples" in terms of "three kids." The leap to "three" will come later.

You may also object that instead of passing out the apples and courting disaster, we could have simply counted them. This is sound advice; thanks. But from the mathematical point of view, is counting the apples really that much different from passing them out? In effect, by handing the apples out, you assign to each apple an owner; different owners to different apples. In counting, you assign to each apple a symbol, which is a string of sounds or letters; again different symbols to different apples. The particular symbols are a matter of convention; Americans say "one, two, three," Germans prefer "eins, zwei, drei," and the French have a taste for "un, deux, trois." The underlying principle is always the same: We assign, in a one-to-one manner, elements of another set to our apples. If all the natural numbers from one to three, and only those, have been assigned, then we know that there are as many apples as there are elements in the set $\{one, two, three\}$. If John, Judy and Jeff each got an apple, and no apple is left or has been passed out to someone else, then we know that there are as many apples as there are kids in the set $\{$*John, Judy, Jeff*$\}$.

Let us call two sets A and B *equipotent* if there are as many elements in A as there are in B. The next definition sums up what you learned in kindergarten.[1]

Definition 1: We say that two sets A and B are *equipotent* and write $A \approx B$ if there exists a one-to-one function $f : A \rightarrow B$ from A onto B.

EXERCISE 1(G): Let A, B, C be sets. Show that:
(a) $A \approx A$.
(b) If $A \approx B$, then $B \approx A$.
(c) If $A \approx B$ and $B \approx C$, then $A \approx C$.

Thus, the notion of equipotency behaves like an equivalence relation. But if $\approx$ were to be a relation, it would have to be a relation on some set $\mathbf{V}$. The only natural candidate for $\mathbf{V}$ is the collection of all sets. We have seen in the Introduction that this collection is too big to form a set; it is a proper class. We already encountered a similar problem in the previous chapter when we discussed order-isomorphism. As in the case of order-isomorphism, you will get the right intuitions if you call $\approx$ an "equivalence" and treat it as if it were an equivalence relation, keeping in the back of your mind that it is a proper class. A formal justification of this approach will be given in Chapter 8.

What does all this tell us about the nature of "three"? A property that John, Paul, and David have in common is that they are all male. In Chapter 1 we showed that male gender can be understood as an equivalence class of a certain equivalence relation. One can also define the common property of three kids, three apples, and three teddybears as an equivalence class of the equivalence $\approx$. This equivalence class is called the *cardinality* of any of the sets in the class.

This is an entirely reasonable approach. The only problem with it is that it fits awkwardly into the framework of axiomatic set theory, since the equivalence classes are in general proper classes.

EXERCISE 2(G): Show that the collection of all sets that are equipotent to $\{\emptyset\}$ is a proper class.

Curiously enough, the standard way of defining cardinal numbers in axiomatic set theory is much closer to the procedure of ordinary counting than it is to the more sophisticated treatment of cardinalities as equivalence classes. In axiomatic set theory, one three-element set is picked somewhat arbitrarily to represent the cardinal number "three." The only difference between set theorists and ordinary mortals is that the latter prefer the set $\{one, two, three\}$, while the former have settled for $\{zero, one, two\}$.[2]

Nevertheless, let us stick to the idea that the cardinality of a set X, which we shall denote by $|X|$, is a kind of property of this set. Mathematically, a "property" can be understood as the collection of all sets Y who share it. In our case, $|X|$ would be the collection of all sets Y such that $X \approx Y$. Cardinal numbers play the

[1]We are not claiming that children really grasp the number concept by the thought process sketched here. If you want to learn more about children's psychology, the writings of Jean Piaget may be of interest to you. We took you on a tour back to childhood only because we think that the concept of cardinality can be best understood if one temporarily suspends all previous knowledge about numbers.

[2]In the next chapter, we shall see which sets correspond to these words.

role of names for these properties. In axiomatic set theory, the names should be sets themselves. More precisely, a name for the property $|X|$ will be a fixed set Y such that $X \approx Y$.[3] Cardinals will be denoted by lower case German (Fraktur) letters: $\mathfrak{m}$, $\mathfrak{n}$, $\mathfrak{p}$, etc.

Definition 2: Let A, B be sets. We write $|A| \leq |B|$ and say that *the cardinality of A is less than or equal to the cardinality of B* if there exists a one-to-one function $f : A \to B$.

EXERCISE 3(G): Show that the notion $\leq$ is well defined. That is, show that if $|A| = |A'|$ and $|B| = |B'|$, then there exists a one-to-one function from A into B iff there exists a one-to-one function from A' into B'.

Note that if $A \subseteq B$, then $|A| \leq |B|$. As our terminology suggests, the notion $\leq$ behaves like a partial order relation on the class of all cardinals. But we shall see in Chapter 11 that the collection of all cardinals is a proper class. Thus, all the admonitions concerning the use of the equivalences $\cong$ and $\approx$ apply to $\leq$ as well. We shall refer to $\leq$ as a "partial ordering," which we take as meaning the same as "partial order relation," except that it may be defined on a proper class.

Let us now consider the properties of $\leq$ in more detail. Reflexivity and transitivity are easy.

EXERCISE 4(G): Prove that for all sets A, B, C the following hold:
(a) $|A| \leq |A|$.
(b) If $|A| \leq |B|$ and $|B| \leq |C|$, then $|A| \leq |C|$.

How about antisymmetry? This is a deep theorem.

Theorem 3 (The Cantor-Schröder-Bernstein Theorem): *For any sets A, B, if $|A| \leq |B|$ and $|B| \leq |A|$, then $|A| = |B|$.*

In most textbooks, you will find a slightly different formulation of this theorem. Here it is.

Theorem 3′ (The Cantor-Schröder-Bernstein Theorem): *Let A, B be sets. If there exist both a one-to-one function $f : A \to B$ and a one-to-one function $g : B \to A$, then there exists a one-to-one function $h : A \to B$ that maps A onto B.*

We shall prove Theorem 3′ in the next chapter. For now, you just need to

[3]In Volume I of this book, we shall adhere to the following terminology: Whenever we use the word "cardinality," we want you to think of it as an equivalence class of the equivalence $\approx$. The phrase "cardinal numbers" will be used exclusively for the representatives, or names, of cardinalities. They will be formally defined in Chapter 11. When we want to be a bit vague, we use the word "cardinal." This can be interpreted either way. For instance, we shall soon consider sets of cardinals. Since a proper class can never be an element of a set, these sets must be composed of cardinal numbers. But when working with an individual cardinal from this set, it may be more enlightening to conceive of it as a cardinality. It should be pointed out that our distinction between "cardinality," "cardinal," and "cardinal number" is merely a pedagogical device to help you become fluent in these concepts. Most authors (including ourselves in Volume II) use these expressions interchangeably.

convince yourself that Theorem 3′ says the same thing as Theorem 3. Later in this chapter you will see some examples of how Theorem 3 can be used to compute cardinalities.

Are every two cardinals comparable by $\leq$? The answer is "Yes" in our standard framework.[4] But the proof requires even more advanced techniques than the proof of Theorem 3. We shall discuss it in Chapter 9.

The next theorem gives an alternative characterization of the ordering $\leq$.

Theorem 4: *Let A, B be nonempty sets. Then $|A| \leq |B|$ iff there exists a function g that maps B onto A.*

EXERCISE 5(PG): Prove Theorem 4.

We shall examine the proof of the "if"–direction in some detail in Chapter 9; so you may want to keep your proof for later reference.

We have been talking quite a bit about abstract cardinalities; it is time to look at some examples. Cardinalities of finite sets are the natural numbers: zero, one, two, three, ... As we already indicated, in the next section we shall redefine these familiar objects in terms of sets. The cardinal $|\mathbb{N}|$ is denoted by $\aleph_0$. (Read: "aleph zero". Aleph is the first letter of the Hebrew alphabet.) Instead of an expression like $|\mathbb{N}| \leq |A|$ we shall write $\aleph_0 \leq |A|$. A set of cardinality $\aleph_0$ is also said to be *denumerable.* Note that a set A is denumerable if and only if its elements can be arranged in a sequence $\langle a_i : i \in \mathbb{N} \rangle$ such that every element of the set appears exactly once in the sequence. Such a sequence will be called a *one-to-one enumeration of* A. If some elements of A are possibly repeated, but all of them appear in the sequence, then we call $\langle a_i : i \in \mathbb{N} \rangle$ simply an *enumeration of* A. We shall often write $(a_i)_{i \in \mathbb{N}}$ instead of $\langle a_i : i \in \mathbb{N} \rangle$. A set that is either finite or denumerable is called *countable.* All other sets are *uncountable.* As we mentioned in the introduction, Cantor's discovery that infinity comes in different sizes marked the birth of set theory. Now we present the proof of this famous result.

Theorem 5 (Cantor): *Let A be any set. Then $|A| < |\mathcal{P}(A)|$.*

Proof: If $A = \emptyset$, then $\mathcal{P}(A) = \{\emptyset\}$. Thus, $|A| = 0 < 1 = |\mathcal{P}(A)|$. If $A \neq \emptyset$, then we define a function $F : A \to \mathcal{P}(A)$ by $F(a) = \{a\}$. It is easy to see that this is a one-to-one function. We conclude that $|A| \leq |\mathcal{P}(A)|$.

It remains to show that $|A| \neq |\mathcal{P}(A)|$. Let $F : A \to \mathcal{P}(A)$, and denote $B = \{x \in A : x \notin F(x)\}$. Suppose $B = F(y)$ for some $y \in A$. If $y \in B$, then $y \in F(y)$. Thus, according to B's definition, y is not among the x's that are members of B, and hence $y \notin B$. But if $y \notin B$, then $y \notin F(y)$; and thus y is among the x's that comprise B, i.e., $y \in B$. Either way, we get a contradiction. Therefore, the assumption that $B = F(y)$ for some $y \in A$ must have been false. In other words, F does not map A onto $\mathcal{P}(A)$. Since F was arbitrary, this shows that $|A| \neq |\mathcal{P}(A)|$. □

If you see it for the first time, the reasoning in the second part of the proof of Theorem 5 may look to you like a trick of stage magic played on unsuspecting

---

[4]The proof requires the Axiom of Choice.

graduate students.[5] But it is actually one of the most important proof techniques in set theory. It is often called "Cantor's diagonalization technique" or "Cantor's diagonalization argument." To give you a better feel for the technique, and also to show you where the name comes from, let us examine the construction in a special case. First we need a technical fact.

Claim 6: *Let A be any set. Then* $|\mathcal{P}(A)| = |{}^{A}\{0,1\}|$.

Proof: Recall that ${}^{A}\{0,1\}$ is the set of all functions from A into $\{0,1\}$. Define a function $c : \mathcal{P}(A) \to {}^{A}\{0,1\}$ by:

$$c(X)(a) = \begin{cases} 1 & \text{if } a \in X; \\ 0 & \text{if } a \notin X. \end{cases}$$

EXERCISE 6(G): Show that c is a one-to-one function that maps $\mathcal{P}(A)$ onto ${}^{A}\{0,1\}$. □

The function $c(X)$ is called the *characteristic function of* X and is denoted by χ_X.[6]

Note that it follows from Theorem 5 and Claim 6 that $|A| < |{}^{A}\{0,1\}|$ for all sets A. Let us see how the second part of the proof of Theorem 5 looks in the special case $A = \mathbb{N}$. That is, let us consider any $F : \mathbb{N} \to {}^{\mathbb{N}}\{0,1\}$. We want to show that F is not onto. We can represent F by an infinite matrix $(a_{i,j})_{i,j\in\mathbb{N}}$, where $a_{i,j} = F(i)(j)$. For example, if $F(0) = \chi_{\emptyset}$, $F(1) = \chi_{\mathbb{N}}$, $F(2) = \chi_E$ and $F(3) = \chi_O$ (where E denotes the set of even numbers, O the set of odd numbers), then the upper left corner of our matrix looks like this:

$$\begin{pmatrix} 0 & 0 & 0 & 0 & \cdots \\ 1 & 1 & 1 & 1 & \cdots \\ 1 & 0 & 1 & 0 & \cdots \\ 0 & 1 & 0 & 1 & \cdots \\ \cdots & \cdots & \cdots & \cdots & \cdots \end{pmatrix}$$

Unfortunately, the publisher did not allow us to print the whole matrix.

Let us now change every entry on the *diagonal* of this matrix into its opposite; i.e., let us replace each entry $a_{i,i}$ by $1 - a_{i,i}$. The resulting matrix $(b_{i,j})_{i,j\in\mathbb{N}}$ looks like this:

$$\begin{pmatrix} 1 & 0 & 0 & 0 & \cdots \\ 1 & 0 & 1 & 1 & \cdots \\ 1 & 0 & 0 & 0 & \cdots \\ 0 & 1 & 0 & 0 & \cdots \\ \cdots & \cdots & \cdots & \cdots & \cdots \end{pmatrix}$$

The entries on the diagonal represent the function $g \in {}^{\mathbb{N}}\{0,1\}$ defined by $g(i) = b_{i,i} = 1 - a_{i,i}$ for each $i \in \mathbb{N}$. Suppose g is in the range of F. Then $g = F(i)$ for some $i \in \mathbb{N}$. But $F(i)(i) = a_{i,i} = 1 - g(i)$, which shows that these two functions are not equal. The construction of g is nothing else but building a function that differs from each function $F(i)$ for at least one argument.

---

[5]Of course, you have already seen Russell's Paradox, which involves the same trick.

[6]This popular terminology is ambiguous. It should be used only if the set A is fixed.

How is this construction related to the second half of the proof of Theorem 5? Think about $F(i)$ as the characteristic function χ_{A_i} of a set $A_i \subseteq \mathbb{N}$. Also, $g = \chi_B$ for some set B of natural numbers.

EXERCISE 7(G): Convince yourself that $B = \{i \in \mathbb{N} : i \notin A_i\}$.

The only difference between the present argument and the proof of Theorem 5 is that the linear ordering of $\mathbb{N}$ helped us in visualizing a "diagonal."

Theorem 5 implies that there are infinitely many different infinite cardinalities. To see this, note that

$$(*) \qquad |\mathbb{N}| < |\mathcal{P}(\mathbb{N})| < |\mathcal{P}(\mathcal{P}(\mathbb{N}))| < \dots .$$

In other words, one can construct an increasing sequence of infinite cardinalities. Are these all the cardinalities of infinite sets? No. The next theorem tells us how to go one step further.

Theorem 7: *Let $\mathcal{M}$ be a set of cardinals such that $\forall \mathfrak{m} \in \mathcal{M} \exists \mathfrak{n} \in \mathcal{M}\,(\mathfrak{m} < \mathfrak{n})$, and let $\mathcal{A}$ be a family of sets such that $\forall \mathfrak{m} \in \mathcal{M} \exists A \in \mathcal{A}\,(|A| = \mathfrak{m})$. Then $|\bigcup \mathcal{A}|$ exceeds every cardinal in $\mathcal{M}$.*

Proof: Let $\mathcal{M}, \mathcal{A}$ be as in the assumption, and suppose toward a contradiction that $|\bigcup \mathcal{A}| \leq \mathfrak{m}$ for some $\mathfrak{m} \in \mathcal{M}$. Pick $\mathfrak{n} \in \mathcal{M}$ and $A, B \in \mathcal{A}$ such that $|B| = \mathfrak{m} < \mathfrak{n} = |A|$. Let $f : \bigcup \mathcal{A} \to B$ be a one-to-one function. Since $A \subseteq \bigcup \mathcal{A}$, we can consider the restriction $h = f|A$ of f to A. This is also a one-to-one function, and it maps A into B. Therefore, $\mathfrak{n} = |A| \leq |B| = \mathfrak{m}$, which contradicts our choice of $\mathfrak{m}$ and $\mathfrak{n}$. $\square$

EXERCISE 8(G): Use Theorem 7 to find a set X of cardinality bigger than that of all the sets appearing in $(*)$.

Once we have found a set X as in Exercise 8, we can produce infinitely many new infinite cardinalities:

$$(**) \qquad |X| < |\mathcal{P}(X)| < |\mathcal{P}(\mathcal{P}(X))| < \dots .$$

Now, we can invoke Theorem 7 to get an even bigger cardinal, and continue the process until ... .

EXERCISE 9(X): Is there a natural end to this process of forming new infinite cardinals? We recommend this exercise instead of counting sheep when you have trouble falling asleep.

What are the cardinalities of the familiar sets $\mathbb{N}, \mathbb{Z}, \mathbb{Q}, \mathbb{P}, \mathbb{R}$ and $\mathbb{C}$? Every real number x has a decimal representation $x = sa_na_{n-1}\dots a_0.b_0b_1b_2\dots$, where $a_i, b_i \in \{0, 1, \dots, 9\}$, $a_n \neq 0$, and $s \in \{+, -\}$. This representation is unique for almost all reals. One notable exception is zero, which can be represented as $-.000\dots$ or $+.000\dots$. The other exceptional case occurs if $b_i = 9$ for all i bigger than a certain j. Such a number could also be represented by changing the rightmost digit c that is different from 9 to $c+1$, and all the 9's to the right of it to 0's. This minor nuisance is usually dealt with by discarding the nonstandard representations: $-.000\dots$ and those ending in an infinite sequence of 9's. With this adjustment being made, there is a one-to-one correspondence between the set $\mathbb{R}$ of reals and the set of standard decimal representations of reals. In this chapter, we shall simply identify each real with its standard decimal representation.

Now it is easy to define a one-to-one function F from ${}^{\mathbb{N}}\{0,1\}$ into $\mathbb{R}$: Just let $F(g) = +.g(0)g(1)g(2)\dots$ The existence of such F proves the inequality $|\mathbb{R}| \geq |{}^{\mathbb{N}}\{0,1\}|$. Thus, by Theorem 5 and Claim 6, the set of reals is uncountable.

Before we pin down the exact size of the set of reals, let us first compare it with some other familiar sets. We have already seen that $|\mathbb{R}| = |(-\frac{\pi}{2}, \frac{\pi}{2})|$. Now let a, b be arbitrary reals such that $a < b$.

EXERCISE 10(G): Show that $|\mathbb{R}| = |(a, b)|$.

How about the closed interval? Since $[a, b] \subset \mathbb{R}$, we must have $|[a, b]| \leq |\mathbb{R}|$. Similarly, $|(a, b)| \leq |[a, b]|$. By Exercise 10, the left hand side of the last inequality is equal to $|\mathbb{R}|$; and now it follows from the Cantor-Schröder-Bernstein Theorem that $|[a, b]| = |\mathbb{R}|$. A similar reasoning shows that also $|[a, b)| = |\mathbb{R}|$ and $|(a, b]| = |\mathbb{R}|$.

EXERCISE 11(PG): If you want to fully appreciate the convenience offered by the Cantor-Schröder-Bernstein Theorem, find an explicit definition of a one-to-one function from $\mathbb{R}$ onto $[0, 1]$.

EXERCISE 12(G): Show that $|\mathbb{C}| = |[0, 1) \times [0, 1)|$.

After having seen that $\mathbb{N}$ and $\mathbb{R}$ are of different size, it may come as a bit of a surprise that there are as many reals as points in the complex plane $\mathbb{C}$.

Theorem 8: $|\mathbb{C}| = |\mathbb{R}|$.

Proof: By Exercise 12, we may as well prove that $[0, 1) \approx [0, 1) \times [0, 1)$. Since $[0, 1) \approx [0, 1) \times \{0\}$, and $[0, 1) \times \{0\} \subset [0, 1) \times [0, 1)$, we have

$$|[0, 1)| \leq |[0, 1) \times [0, 1)|.$$

By the Cantor-Schröder-Bernstein Theorem, the proof of Theorem 8 boils down to finding a one-to-one function $G : [0, 1) \times [0, 1) \to [0, 1)$. So suppose $\langle x, y\rangle \in [0, 1) \times [0, 1)$, and let $x = +.x_0x_1x_2\dots$, $y = +.y_0y_1y_2\dots$ Define $G(x, y) = z$, where $z = +.x_0y_0x_1y_1x_2y_2\dots$.

EXERCISE 13(G): Convince yourself that G is indeed one-to-one. □

Note that the function G does *not* map $[0, 1) \times [0, 1)$ onto $[0, 1)$.

EXERCISE 14(PG): Why?

Again, the Cantor-Schröder-Bernstein Theorem made life much easier. Here is another application of it.

Theorem 9: $|\mathbb{R}| = |{}^{\mathbb{N}}\{0,1\}|$.

Proof: Since we have already shown that $|\mathbb{R}| \geq |{}^{\mathbb{N}}\{0,1\}|$, it suffices to prove the converse inequality. By Theorem 4 and Exercise 10, it suffices to find a function H that maps the set ${}^{\mathbb{N}}\{0,1\}$ onto $[0,1]$. For $f \in {}^{\mathbb{N}}\{0,1\}$, define

$$H(f) = \sum_{n \in \mathbb{N}} f(n) \cdot 2^{-(n+1)}.$$

EXERCISE 15(PG): Show that the function H maps ${}^{\mathbb{N}}\{0,1\}$ onto $[0,1]$. □

How big are the sets $\mathbb{Z}$ of integers and $\mathbb{Q}$ of rationals? It is not hard to construct a one-to-one function f from $\mathbb{Z}$ onto $\mathbb{N}$: For each nonnegative integer n, let $f(n) = 2n$. For each negative integer m, let $f(m) = 2|m| - 1$.

EXERCISE 16(G): Prove that the function f is one-to-one and maps $\mathbb{Z}$ onto $\mathbb{N}$.

It follows that $\mathbb{Z}$ is a denumerable set. This argument can easily be generalized.

EXERCISE 17(G): (a) Show that if A, B are disjoint denumerable sets, then $A \cup B$ is a denumerable set.

(b) The same as (a), but *without* the assumption that $A \cap B = \emptyset$. You may use any theorems presented so far in this section.

One can generalize even further.

Definition 10: Let $\mathfrak{m}, \mathfrak{n}$ be cardinals, and let A, B be disjoint sets such that $|A| = \mathfrak{m}$ and $|B| = \mathfrak{n}$. Define: $\mathfrak{m} + \mathfrak{n} = |A \cup B|$. [7]

EXERCISE 18(G): (a) Show that "+" is a well defined operation on cardinals.

(b) Convince yourself that for finite cardinals, the operation "+" is precisely the familiar addition of natural numbers.

In our new terminology, the result of Exercise 17(a) can be expressed as follows.

Theorem 11: $\aleph_0 + \aleph_0 = \aleph_0$.

Let us now generalize the operations of multiplication and exponentiation of natural numbers to operations on arbitrary cardinals.

Definition 12: Let $\mathfrak{m}$, $\mathfrak{n}$ be cardinals, and let A, B be sets such that $|A| = \mathfrak{m}$ and $|B| = \mathfrak{n}$. We define:

(a) $\mathfrak{m} \cdot \mathfrak{n} = |A \times B|$;

(b) $\mathfrak{m}^{\mathfrak{n}} = |{}^{B}A|$.

EXERCISE 19(G): (a) Show that the operations $\mathfrak{m} \cdot \mathfrak{n}$ and $\mathfrak{m}^{\mathfrak{n}}$ are well defined.

---

[7] Alternatively, if one does not want to assume that A and B are disjoint, one can "tag" them by zero and one and define $\mathfrak{m} + \mathfrak{n} = |A \times \{0\} \cup B \times \{1\}|$.

(b) Convince yourself that if $\mathfrak{m}$ and $\mathfrak{n}$ are finite cardinals, then the operations defined above coincide with the familiar multiplication and exponentiation of natural numbers.

There is one exception to the claim made in Exercise 19(b). Note that ${}^{\emptyset}\emptyset = \{\emptyset\}$, since the empty set is a function with empty domain and empty range. Therefore, in cardinal arithmetic, $0^0 = |{}^{\emptyset}\emptyset| = 1$. Recall that in calculus the number 0^0 is not defined; it is treated as an indeterminate form.

Many properties of multiplication and exponentiation of natural numbers remain true for infinite cardinals.

Theorem 13: *Let* $\mathfrak{m}, \mathfrak{n}, \mathfrak{p}, \mathfrak{r}$ *be cardinals. Then:*
(a) $\mathfrak{m} \cdot \mathfrak{n} = \mathfrak{n} \cdot \mathfrak{m}$;
(b) $(\mathfrak{m} \cdot \mathfrak{n}) \cdot \mathfrak{p} = \mathfrak{m} \cdot (\mathfrak{n} \cdot \mathfrak{p})$;
(c) $(\mathfrak{m} + \mathfrak{n}) \cdot \mathfrak{p} = (\mathfrak{m} \cdot \mathfrak{p}) + (\mathfrak{n} \cdot \mathfrak{p})$;
(d) $\mathfrak{m}^{\mathfrak{n}+\mathfrak{p}} = (\mathfrak{m}^{\mathfrak{n}}) \cdot (\mathfrak{m}^{\mathfrak{p}})$;
(e) $(\mathfrak{m}^{\mathfrak{n}})^{\mathfrak{p}} = \mathfrak{m}^{\mathfrak{n} \cdot \mathfrak{p}}$;
(f) *if* $\mathfrak{m} \leq \mathfrak{n}$ *and* $\mathfrak{p} \leq \mathfrak{r}$, *then:*
  (f1) $\mathfrak{m} + \mathfrak{p} \leq \mathfrak{n} + \mathfrak{r}$,
  (f2) $\mathfrak{m} \cdot \mathfrak{p} \leq \mathfrak{n} \cdot \mathfrak{r}$,
  (f3) *if* $\mathfrak{n} > 0$, *then* $\mathfrak{m}^{\mathfrak{p}} \leq \mathfrak{n}^{\mathfrak{r}}$.

Proof: As an illustration, we prove (e). Let A, B, C be such that $|A| = \mathfrak{m}$, $|B| = \mathfrak{n}$, and $|C| = \mathfrak{p}$. We want to show that $|{}^{B\times C}A| = |{}^{C}({}^{B}A)|$. In other words, we want to find a bijection $F : {}^{B\times C}A \longrightarrow {}^{C}({}^{B}A)$. This will not require any brilliant ideas; we must only carefully keep track of the nature of the sets we are dealing with.

Let $f \in {}^{B\times C}A$. We want to assign to f a function $F(f) = g$ such that $g : C \longrightarrow {}^{B}A$. This means that each value $g(c)$ will also be a function; this time a function from B into A. Thus, we must assign a function value $g(c)(b) \in A$ to each $b \in B$. The most natural choice is $F(f)(c)(b) = f(\langle b, c\rangle)$. This defines a function F.

Let us see whether this function is a bijection. If $f_0 \neq f_1$, then there is a pair $\langle b, c\rangle \in B \times C$ such that $f_0(\langle b, c\rangle) \neq f_1(\langle b, c\rangle)$. Our definition implies that the functions $F(f_0)(c)$ and $F(f_1)(c)$ differ on the argument b, and thus the functions $F(f_0)$ and $F(f_1)$ differ on the argument c. Hence, $F(f_0)$ and $F(f_1)$ are different functions; and we have shown that F is an injection. To show that it is also a surjection, let $g \in {}^{C}({}^{B}A)$. Define $f : B \times C \longrightarrow A$ by: $f(\langle b, c\rangle) = g(c)(b)$. It is not hard to verify that $F(f) = g$. Thus, every element of ${}^{C}({}^{B}A)$ is in the range of F, which means that F is a surjection. □

Since keeping track of the nature of the sets involved was the whole point of the above proof, we dutifully wrote $f(\langle b, c\rangle)$. This manner of writing is usually considered to be overly compulsive. In the remainder of this text, we shall use the more informal notation $f(b, c)$. Did you notice that this more relaxed style has already been used a few pages earlier?

EXERCISE 20(PG): Prove the remaining parts of Theorem 13.

Now we prove the analogue of Theorem 11 for cardinal multiplication.

Theorem 14: $\aleph_0 \cdot \aleph_0 = \aleph_0$.

Proof: We want to find a one-to-one enumeration $\langle\langle n_i, m_i\rangle : i \in \mathbb{N}\rangle$. Here is one: Start with $\langle n_0, m_0\rangle = \langle 0, 0\rangle$.

Next, list all pairs $\langle n, m\rangle$ with $n + m = 1$ in increasing order of the first coordinates: $\langle n_1, m_1\rangle = \langle 0, 1\rangle$, $\langle n_2, m_2\rangle = \langle 1, 0\rangle$.

Next, list all pairs with $n+m = 2$ in a similar way: $\langle n_3, m_3\rangle = \langle 0, 2\rangle$, $\langle n_4, m_4\rangle = \langle 1, 1\rangle$, $\langle n_5, m_5\rangle = \langle 2, 0\rangle$.

And so on. This idea can also be nicely shown in a picture:

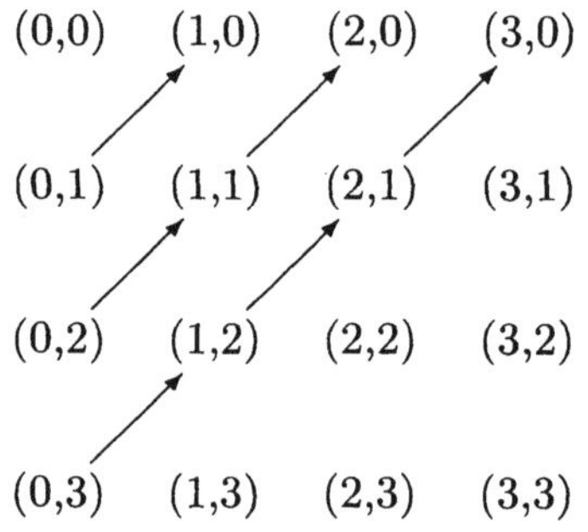

It is easy to *see* that the function $I : \mathbb{N} \to \mathbb{N} \times \mathbb{N}$ defined by $I(i) = \langle n_i, m_i\rangle$ is a bijection. In order to *prove* this, we would like to define this function by a formula rather than by a picture. It is actually easier to derive a concise formula for its inverse function $J : \mathbb{N} \times \mathbb{N} \to \mathbb{N}$.

The first step in finding a formula for J is the observation that $J(n, m)$ is the number of pairs $\langle n', m'\rangle$ such that $n' + m' < n + m$, or $n' + m' = n + m$ and $n' < n$. In other words, if $s(\ell)$ is the number of pairs $\langle n', m'\rangle$ with $n' + m' = \ell$, then

$$J(n, m) = n + \sum_{\ell=0}^{n+m-1} s(\ell).$$

Next observe that $s(\ell) = \ell + 1$ for each $\ell \in \mathbb{N}$. We conclude that

$$J(n, m) = n + \sum_{\ell=0}^{n+m-1} (\ell + 1) = n + \frac{(n + m)(n + m + 1)}{2}.$$

EXERCISE 21(PG): Prove that J is one-to-one from $\mathbb{N} \times \mathbb{N}$ onto $\mathbb{N}$. □

Corollary 15: $\mathbb{Q} \approx \mathbb{N}$.

Proof: Since $\mathbb{N} \subseteq \mathbb{Q}$, the inequality $|\mathbb{N}| \leq |\mathbb{Q}|$ holds. Hence, by the Cantor-Schröder-Bernstein Theorem, it suffices to show that $|\mathbb{Q}| \leq |\mathbb{N}|$. It follows from Theorem 14 that $|\mathbb{Z} \times \mathbb{N}| = |\mathbb{N}|$. Therefore, all we need is a one-to-one function $f : \mathbb{Q} \to \mathbb{Z} \times \mathbb{N}$. We can get one by letting $f(0) = \langle 0, 1\rangle$, and assigning to each $q \in \mathbb{Q}\backslash\{0\}$ a pair

$$\langle n, m\rangle \in (\mathbb{Z}\backslash\{0\}) \times (\mathbb{N}\backslash\{0\})$$

such that $|n|$ and $|m|$ are relatively prime, and $q = \frac{n}{m}$.

EXERCISE 22(G): Convince yourself that this function f is well defined and one-to-one. □

Corollary 16: *If n is a positive natural number, then $|\mathbb{N}^n| = |\mathbb{Z}^n| = |\mathbb{Q}^n| = \aleph_0$.*

The proof of Corollary 16 will be given in the next section.

Theorem 17: *Let $\mathcal{A}$ be a denumerable family of sets.*
(a) *If each set in $\mathcal{A}$ is denumerable, then so is $\bigcup \mathcal{A}$.*
(b) *If each set in $\mathcal{A}$ is countable, then so is $\bigcup \mathcal{A}$.*

Proof: (a) Let $\mathcal{A} = \{A_i : i \in \mathbb{N}\}$. Since $A_0 \subseteq \bigcup \mathcal{A}$ and $|A_0| = \aleph_0$, the inequality $\aleph_0 \leq |\bigcup \mathcal{A}|$ holds. It remains to show that $\aleph_0 \geq |\bigcup \mathcal{A}|$. We do this by finding a function f that maps $\mathbb{N} \times \mathbb{N}$ onto $\bigcup \mathcal{A}$. The desired cardinal inequality then follows from Theorems 4 and 14. Choose for each $i \in \mathbb{N}$ an enumeration $\langle a^i_j : j \in \mathbb{N} \rangle$ of the set A_i, and let $f(i,j) = a^i_j$. The verification that this function maps $\mathbb{N} \times \mathbb{N}$ onto $\bigcup \mathcal{A}$ is so easy that we do not even call it an exercise.

For the proof of (b), let $\mathcal{A}$ be a denumerable family of countable sets, and let $\mathcal{B} = \{A \cup \mathbb{N} : A \in \mathcal{A}\}$. This is a denumerable family of denumerable sets, and hence $\bigcup \mathcal{B}$ is denumerable by point (a). Clearly, $\bigcup \mathcal{A} \subseteq \bigcup \mathcal{B}$. It may sound obvious to you that any subset of a denumerable set is countable; and in the next chapter we are actually going to prove this fact. So, in particular, $\bigcup \mathcal{A}$ is countable. □

Now we give an interesting application of Theorem 17. Recall that a real number x is *algebraic* if it is the root of a polynomial with integer coefficients. In other words, x is algebraic if there are a natural number n and integers $a_0, \dots, a_n$ with $a_n \neq 0$ such that $a_n x^n + a_{n-1} x^{n-1} + \dots + a_1 x^1 + a_0 = 0$. For example, each rational number $\frac{k}{m}$ is a root of the polynomial $mx - k$, and the irrational $\sqrt{2}$ is a root of $x^2 - 2$. Let us denote the set of all algebraic numbers by $\mathbb{A}$.

Theorem 18: $|\mathbb{A}| = \aleph_0$.

Proof: For each $n \in \mathbb{N}$, let P_n be the set of polynomials of degree n with integer coefficients. We assign to each polynomial $p = a_n x^n + a_{n-1} x^{n-1} + \dots + a_1 x^1 + a_0 \in P_n$ an $n+1$–tuple $f_n(p) = \langle a_n, a_{n-1}, \dots, a_0 \rangle \in \mathbb{Z}^{n+1}$. Note that f_n is a one-to-one function from P_n onto the denumerable set $(\mathbb{Z} \backslash \{0\}) \times \mathbb{Z}^n$. Thus, each set P_n is denumerable. For each $p \in P_n$, let $R(p) = \{x \in \mathbb{R} : p(x) = 0\}$ be the set of real roots of p. It follows from the Fundamental Theorem of Algebra that $|R(p)| \leq n$ for each $p \in P_n$. For a fixed n, let $\mathbb{A}_n = \bigcup \{R(p) : p \in P_n\}$. This is the union of a denumerable family of countable sets. By Theorem 17(b), for each n the set $\mathbb{A}_n$ is countable. For the same reason, the union $\bigcup_{n \in \mathbb{N}} \mathbb{A}_n$ is also countable. The latter union is easily recognized as the set $\mathbb{A}$ of algebraic reals. To see that $\mathbb{A}$ is denumerable rather than just countable, recall that $\mathbb{A}$ contains the infinite set $\mathbb{Q}$, and hence, $\mathbb{A}$ itself must be infinite. □

EXERCISE 23(G): Prove that the family Fin of finite subsets of $\mathbb{N}$ is denumerable.

EXERCISE 24(PG): Ten years from now, the computer language MAGIC FERMAT[8] will be widely used. This language is currently being designed especially for the purpose of computing functions on $\mathbb{N}$ that take integer values. You will be able to specify algorithms for computing such functions by encoding the algorithm in a finite sequence of MAGIC FERMAT instructions. Supposedly, MAGIC FERMAT instructions will be radically different from those used in contemporary computer languages. Details are top secret, but confidential leaks from well informed sources indicate that the set of MAGIC FERMAT instructions will be finite. Prove that the set of functions from $\mathbb{N}$ into $\mathbb{Z}$ that can be computed by a MAGIC FERMAT algorithm is countable.

Theorems 11 and 14 have analogues for larger infinite cardinals.

Theorem 19: *Let* $\mathfrak{m}, \mathfrak{n}$ *be cardinals such that* $\max\{\mathfrak{m}, \mathfrak{n}\} \geq \aleph_0$. *Then* $\mathfrak{m} + \mathfrak{n} = \max\{\mathfrak{m}, \mathfrak{n}\}$. *If, moreover,* $\mathfrak{m}, \mathfrak{n} > 0$, *then* $\mathfrak{m} \cdot \mathfrak{n} = \max\{\mathfrak{m}, \mathfrak{n}\}$.

The part of Theorem 19 that pertains to multiplication of cardinals is called *Hessenberg's Theorem.* As we mentioned earlier, every two cardinals are comparable by $\leq$. Thus, the use of "max" in Theorem 19 makes sense. We shall prove this theorem in Chapter 11.

Corollary 20: *Let* $\mathbb{P}$ *be the set of irrationals. Then* $|\mathbb{P}| = |\mathbb{R}|$.

Proof: Since $\mathbb{P} \cup \mathbb{Q} = \mathbb{R}$ and $\mathbb{P} \cap \mathbb{Q} = \emptyset$, it follows from Definition 10 that $|\mathbb{P}| + |\mathbb{Q}| = |\mathbb{R}|$. By Theorem 19, $|\mathbb{R}| = \max\{|\mathbb{P}|, |\mathbb{Q}|\}$. Since $|\mathbb{Q}| \neq |\mathbb{R}|$, it must be the case that $|\mathbb{P}| = |\mathbb{R}|$. □

EXERCISE 25(G): Let $\mathbb{T}$ be the set of *transcendental* real numbers, i.e., the set of those reals that are not algebraic. Show that $|\mathbb{T}| = |\mathbb{R}|$.

So far, we have avoided assigning a symbol to $|\mathbb{R}|$. In the early years of set theory, the notation $|\mathbb{R}| = \mathfrak{c}$ was popular. The letter $\mathfrak{c}$ stands for "continuum", a word that we shall occasionally use to designate the cardinality of $\mathbb{R}$. We shall not use the symbol $\mathfrak{c}$ much, because this notation gives absolutely no information about how $|\mathbb{R}|$ is related to other cardinals. Since $|\mathbb{R}| = |{}^{\mathbb{N}}\{0,1\}|$, we have $|\mathbb{R}| = 2^{\aleph_0}$; and this notation will be used in most of this book. Still, this notation does not tell us exactly where the continuum fits into the class of all cardinals. A natural question is whether $2^{\aleph_0}$ is *the smallest* uncountable cardinal. Cantor conjectured that this is indeed the case. His conjecture is called the *Continuum Hypothesis* and abbreviated by CH.

Continuum Hypothesis (CH): *Every subset of the set of reals is either countable or of cardinality* $\mathfrak{c}$.

EXERCISE 26(G): Prove that the version of CH given above says indeed that $2^{\aleph_0}$ is the smallest uncountable cardinal. You may take it for granted that every two cardinals are comparable.

[8]Unofficial trademark of Just & Weese, Ltd.

In order to see that CH is a reasonable conjecture, recall that all subsets of $\mathbb{R}$ considered so far are indeed either countable or of size $\mathfrak{c}$. On the other hand, each of the sets $\mathbb{N}$, $\mathbb{Z}$, $\mathbb{Q}$, $\mathbb{P}$, $\mathbb{A}$, $\mathbb{T}$, (a, b), $[a, b)$, $[a, b]$ has a very simple definition.[9] Couldn't there be some "wild" uncountable subset of $\mathbb{R}$ of cardinality strictly less than $\mathfrak{c}$?

Over many decades, this problem held a fascination for set theorists similar to the one the Riemann Hypothesis holds for analysts. Cantor spent much of his time trying to prove the Continuum Hypothesis. In the famous address given at the International Congress of Mathematicians in Paris in 1900, David Hilbert put the Continuum Hypothesis as Problem Number One on a list of problems which he considered the potentially most stimulating ones for the mathematics of the 20th century. History confirmed most of Hilbert's intuitions; the problems on his list led indeed to major developments.

In 1938, Kurt Gödel proved that it is impossible to *disprove* the Continuum Hypothesis without destroying the foundations of contemporary mathematics. The most remarkable feature of Gödel's result is that he *did not prove* CH. We shall see in Chapter 6 how it is, in principle, possible to show that a statement cannot be disproved without actually proving it. In Chapter 12, we shall give a broad outline of Gödel's proof. The proof is the first construction of an *inner model of set theory.* Inner model theory is nowadays a rapidly developing branch of set theory.

In 1963, Paul J. Cohen showed that it is impossible to *prove* CH without destroying the foundations of contemporary mathematics. This settles the problem by showing that the truth or falsity of CH is beyond our present cognitive means. In Chapter 6, we shall explain the nature of such results. For his proof, Cohen invented a new method called the "forcing method." Since 1963, this method has been used by many mathematicians to prove literally hundreds of theorems.

It follows from the above that while addition and multiplication of infinite cardinals are a trivial matter, the rules governing exponentiation of infinite cardinals are somewhat elusive. Chapter 11 contains a detailed account of what is known about the latter operation.

We conclude the present chapter with computations of the cardinalities of some sets of functions.

Let us start with the set ${}^{\mathbb{R}}\{0,1\}$. By definition, its cardinality is $2^{2^{\aleph_0}}$. How does $|{}^{\mathbb{R}}\{0,1\}|$ compare with $|{}^{\mathbb{R}}\mathbb{R}|$? Of course, since ${}^{\mathbb{R}}\{0,1\} \subset {}^{\mathbb{R}}\mathbb{R}$, the former cardinal does not exceed the latter. But are they equal or different? Let us make creative use of the theorems of this section. $|{}^{\mathbb{R}}\mathbb{R}| = (2^{\aleph_0})^{2^{\aleph_0}}$. By Theorem 13(e), $(2^{\aleph_0})^{2^{\aleph_0}} = 2^{(\aleph_0 \cdot 2^{\aleph_0})}$. Applying Theorem 19 to the exponent of the right hand side, we get $2^{(\aleph_0 \cdot 2^{\aleph_0})} = 2^{2^{\aleph_0}} = |{}^{\mathbb{R}}\{0,1\}|$. It follows that there are as many functions from the real line into $\{0, 1\}$ as there are functions from the real line into the real line.

EXERCISE 27(G): Prove that if $\mathfrak{m}$ is an infinite cardinal, and $\mathfrak{n}$ is a cardinal such that $2 \leq \mathfrak{n} \leq \mathfrak{m}$, then $2^{\mathfrak{m}} = \mathfrak{n}^{\mathfrak{m}}$.

EXERCISE 28(G): Compute $|{}^{\mathbb{Q}}\mathbb{R}|$.

[9] Speaking more scientifically, each of these sets is a *Borel subset of* $\mathbb{R}$. Borel sets will be formally defined in Chapter 13. It can be proved that every Borel subset of the real line is either countable or of cardinality $\mathfrak{c}$.

Finally, let us compute $|C(\mathbb{R},\mathbb{R})|$, where $C(\mathbb{R},\mathbb{R})$ is the set of all *continuous* functions from the set of reals into the set of reals. Since each constant function is continuous, and for each real number y there is a constant function that takes the value y all the time, the inequality $2^{\aleph_0} \leq |C(\mathbb{R},\mathbb{R})|$ holds. On the other hand, for each $f \in C(\mathbb{R},\mathbb{R})$, let $G(f) = f|\mathbb{Q}$ be the restriction of f to the rationals. Then $G : C(\mathbb{R},\mathbb{R}) \to {}^{\mathbb{Q}}\mathbb{R}$. Moreover, it is a well known theorem that if two functions in $C(\mathbb{R},\mathbb{R})$ agree on all rationals, then they are equal. In other words, the function G is one-to-one, and we have shown that $|C(\mathbb{R},\mathbb{R})| \leq |{}^{\mathbb{Q}}\mathbb{R}| = 2^{\aleph_0}$. By the Cantor-Schröder-Bernstein Theorem, $|C(\mathbb{R},\mathbb{R})| = 2^{\aleph_0}$.

Mathographical Remarks

Our notation $|X|$ for the cardinality of X is fairly standard, but the notations $\overline{\overline{X}}$ and $\|X\|$ are also common. Theorem 3′ is called the "Cantor-Bernstein Theorem" by 45% of authors, and the "Schröder–Bernstein Theorem" by another 45% . In an attempt to make everybody happy, we side with the minority and use all three names.

The theorem that Borel subsets of the real line are either countable or of cardinality $\mathfrak{c}$ (see footnote 9) belongs to a branch of set theory called *Descriptive Set Theory.* Roughly speaking, descriptive set theorists study connections between properties of sets and the way these sets are defined. It is an exciting, rapidly developing branch of set theory with many applications to other areas of mathematics. Unfortunately, the study of definability requires more prerequisites from logic and topology than we wanted to assume in this book. Therefore, we decided to entirely omit this branch of set theory. The book *Descriptive Set Theory,* by Yiannis Moschovakis, North-Holland, 1980, and the monograph *Classical Descriptive Set Theory,* by Alexander Kechris, Springer-Verlag, 1995, give good introductions to the subject.

CHAPTER 4

Induction

4.1. Induction and recursion over the set of natural numbers

Theorem 1: *If n is a positive natural number, then* $|\mathbb{N}^n| = \aleph_0$.

You already encountered this theorem as part of Corollary 3.16.

Proof of Theorem 1: In this proof we consider $\mathbb{N}^n$ as $\mathbb{N}$ for $n = 1$ and as the set of all n-tuples of natural numbers for $n > 1$. Recall that an n-tuple $\langle a_0, \dots a_{n-1}\rangle$ is $\langle\langle \dots \langle a_0, a_1\rangle, \dots\rangle, a_{n-1}\rangle$, and note that $\mathbb{N}^{n+1} = \mathbb{N}^n \times \mathbb{N}$ for $n \geq 1$.

(I1) If $n = 1$, then $\mathbb{N}^n = \mathbb{N}^1 = \mathbb{N}$, and this set has cardinality $\aleph_0$ by definition.

(I2) Suppose $|\mathbb{N}^k| = \aleph_0$. Then $\mathbb{N}^{k+1} = \mathbb{N}^k \times \mathbb{N}$, and therefore, $|\mathbb{N}^{k+1}| = |\mathbb{N}^k| \cdot |\mathbb{N}|$. By our inductive assumption, $|\mathbb{N}^k| \cdot |\mathbb{N}| = \aleph_0 \cdot |\mathbb{N}|$, and hence $|\mathbb{N}^{k+1}| = \aleph_0 \cdot \aleph_0 = \aleph_0$. The last equation follows from Theorem 3.14.

(I3) Since the equality $|\mathbb{N}^n| = \aleph_0$ holds for $n = 1$, and its truth for $n = k$ implies its truth for $n = k + 1$, it must hold for each positive integer n. □

EXERCISE 1(G): Convince yourself that the remaining parts of Corollary 3.16 can be proved by the same kind of argument.

The proof of Theorem 1 is our first example of a *proof by mathematical induction.* We assume that you are familiar with straightforward inductive proofs like the one of Theorem 1. This chapter is devoted to more sophisticated variants of this technique.

Let us analyze the proof of Theorem 1. It has three parts. The first part (I1) is needed to get started. Since we want to prove Theorem 1 for all *positive* natural numbers, we start with $n = 1$. If you want to prove a theorem for *all* natural numbers, you would typically start with $n = 0$. Sometimes, you start with an even larger n. For example, if you want to show by induction that $3^n > n^3$ for all $n > 3$, then you start with $n = 4$.

The second part (I2) is called the *induction step.* Its first sentence is the *inductive assumption.* In a straightforward inductive proof like that of Theorem 1, the inductive assumption is easy to formulate. But in the more general types of inductive proofs that we are going to discuss in the next chapter, there may be several possibilities for the inductive assumption. Choosing the right one of these is a crucial step in finding the proof. The remainder of (I2) is a proof of Theorem 1 for $n = k + 1$. At one point in this argument, we used the inductive assumption. If you write an inductive proof, and you do *not* use the inductive assumption at any point of the inductive step, then this is a sure indication of one of two possibilities: either your proof is wrong, or it is not really a proof by mathematical induction.

The third part (I3), the conclusion of the argument, is a summary of the method of induction. In proofs written for experienced readers, this part is usually omitted, or replaced by a sentence like: "By the principle of mathematical induction, the theorem holds." In the next chapter, we shall take a closer look at (I3).

Now we give a formal definition of the set X^k of all k-tuples of elements of a set X.

Definition 2: Let X be any set, and let k be a positive integer. Define:

(R1) $X^1 = X$.

(R2) If X^k has already been defined, let $X^{k+1} = X^k \times X$.

(R3) Since X^1 has been explicitly defined by clause (R1), and clause (R2) gives a definition of X^{k+1} from X^k, we have defined X^n for every positive integer n.

Looks familiar? The clauses (R1)–(R3) correspond exactly to parts (I1)–(I3) of the proof of Theorem 1. The difference is that there we had a proof by *induction,* whereas Definition 2 is an example of a definition by *recursion.* Induction and recusion are two sides of the same coin. While the word "induction" is used in connection with proofs, the word "recursion" refers to definitions.[1]

Recursive definitions can be viewed as wrapping up an infinite process in a finite amount of time. The clause (R2) describes the constructions of the sets X^k as if they would have to be carried out one after the other. Carrying out the actual construction of all X^k's would take an infinite amount of time, and could never be finished. But, since this clause completely *describes* the construction, all X^k's are defined the moment we write down (R2).

One can also look at the inductive step (I2) in an inductive proof as infinitely many separate proofs. Since all these proofs look exactly alike, we can write a single description (I2) for them, and thereby complete the proof in finitely many steps.

Definition 2 was written in such a cumbersome way only for later reference. Normally, it would have been written like this:

Definition 2′: Let X be any set. For a positive integer n, we define X^n recursively by:

$$\begin{aligned} X^1 &= X; \\ X^{k+1} &= X^k \times X. \end{aligned}$$

Let us now define the natural numbers by recursion. As we mentioned in Chapter 1, the natural number n will be the set of all its predecessors: $n = \{0, 1, \dots, n-1\}$. Since 0 has no predecessors, this must be the empty set. Here is a formal definition.

Definition 3: For each natural number we define recursively a set that we shall henceforth identify with this number:

$$\begin{aligned} 0 &= \emptyset; \\ k+1 &= k \cup \{k\}. \end{aligned}$$

[1] Frequently, mathematicians speak of "defining something by induction" when they mean a definition by recursion. Depending on your taste, you may consider this language overly sloppy, or you may find our careful distinction between recursion and induction overly picky.

Isn't Definition 3 circular? That is, aren't we defining natural numbers in terms of natural numbers? This depends on whether you adopt a naive or an axiomatic view of set theory. Recall our discussion at the beginning of the previous chapter. If we assume that natural numbers are "really" objects other than sets (e.g., proper equivalence classes of $\approx$), then it makes sense to assign to each natural number n a set $\widetilde{n}$ that could function as a name for this number. Such names are also called *numerals*. It is reasonable to require that $\widetilde{n}$ should be an n-element set. If we want to take this approach, we should perhaps reformulate Definition 3 as follows:

Definition 3′: For each natural number n we define recursively the numeral $\widetilde{n}$ by:

$$\begin{aligned} \widetilde{0} &= \emptyset; \\ \widetilde{k+1} &= \widetilde{k} \cup \{\widetilde{k}\}. \end{aligned}$$

The set $\{\widetilde{n} : n \in \mathbb{N}\}$ is denoted by ω.

EXERCISE 2(G): (a) Show by induction that for each natural number n the numeral $\widetilde{n}$ is equal to $\{\widetilde{0}, \widetilde{1}, \dots, \widetilde{n-1}\}$.

(b) Express $\widetilde{3}$ using only the empty set symbol and braces.

This approach reflects the way most mathematicians understand the expression "$n = \{0, 1, \dots, n-1\}$." However, it is uncommon to make a notational distinction between n and $\widetilde{n}$, and in most of the remainder of this book we shall not make such a distinction. Also, we shall use the notation ω in place of $\mathbb{N}$ from now on.

In the framework of axiomatic set theory that we start developing in Chapter 7, there is no way to eliminate the circularity of Definition 3. In this system, the only kind of objects that we can study are sets, and everything else has to be reinterpreted as a set. In particular, the natural number n *is* the numeral $\{0, 1, \dots, n-1\}$. Thus, the circularity of Definition 3 cannot be explained away by making a distinction between numbers and numerals. In Chapter 10 we shall present a noncircular way of defining natural numbers in axiomatic set theory.

In Chapter 1, we indicated that in most of this book we shall treat a Cartesian product A^n as the set of functions from n into A. We briefly justified this approach for $n = 2$. The next exercise extends this justification to larger n's.

EXERCISE 3(G): For the purpose of this exercise, let A^n be defined as in Definition 2. Prove by induction that if n is a positive integer and A is any set, then $A^n \approx {}^nA$.

The next definition introduces very important concepts.

Definition 4: Let x be any set.

(a) For each $n \in \omega$ we define recursively a set $\bigcup^{(n)} x$ as follows:

$$\begin{aligned} \bigcup^{(0)} x &= x; \\ \bigcup^{(k+1)} x &= \bigcup(\bigcup^{(k)} x). \end{aligned}$$

(b) We define a set $TC(x)$, called the *transitive closure of* x, by $TC(x) = \bigcup\{\bigcup^{(n)} x : n \in \omega\}$.

(c) x is *transitive* iff $TC(x) = x$. We say that x is *hereditarily finite* (*hereditarily countable*) iff $TC(x)$ is finite (countable).

EXERCISE 4(G): (a) For each $n \in \omega$, find $\bigcup^{(n)} 4$.
(b) Show that each $n \in \omega$ is a transitive set.
(c) Show that ω is a transitive set.

EXERCISE 5(G): (a) Convince yourself that the set $\{\omega\}$ is finite, but not hereditarily finite.
(b) Find a set x that is countable, but not hereditarily countable.

Now we are going to present two proofs of the Cantor-Schröder-Bernstein Theorem. One of the proofs involves a recursive definition, the other one doesn't. Quite frequently, it is possible to replace a recursive definition by some other type of argument. The resulting proofs are often shorter and more elegant than the ones involving recursion, but usually they are also less illuminating.

Theorem 5 (Cantor-Schröder-Bernstein): *Let a, b be arbitrary sets. If $|a| \leq |b|$ and $|b| \leq |a|$, then $|a| = |b|$.*

Proof: First we reduce the theorem to a special case.

Lemma 6: *Suppose a, a', a'' are sets such that $a'' \subseteq a' \subseteq a$ and $a \approx a''$. Then $a \approx a'$.*

To see how Theorem 5 follows from Lemma 6, let a and b be as in the assumptions of Theorem 5, and let $f_1 : a \longrightarrow b$, $g_1 : b \longrightarrow a$ be one-to-one functions. Denote

$$\begin{aligned} a' &= g_1[b]; \\ a'' &= g_1[f_1[a]]. \end{aligned}$$

Then $a'' \subseteq a' \subseteq a$ and $g_1 \circ f_1$ is a one-to-one function from a onto a''. Hence, $a \approx a''$, and the assumptions of Lemma 6 are satisfied. By the lemma, $a \approx a'$. On the other hand, $b \approx g_1[b] = a'$, and we conclude that $a \approx b$.

Proof of Lemma 6: Let a, a', a'' be as in the assumptions of Lemma 6, and let $f : a \longrightarrow a''$ be a one-to-one function from a onto a''.

Claim 7: *There exists $c^* \subseteq a''$ such that $f[c^* \cup (a \backslash a')] = c^*$.*

We show how Claim 7 implies Lemma 6. Let c^* be as in the claim. Define a function $h : a \longrightarrow a'$ by

$$h(x) = \begin{cases} f(x) & \text{if } x \in c^* \cup (a \backslash a'), \\ x & \text{otherwise.} \end{cases}$$

EXERCISE 6(G): Show that h is a one-to-one function from a onto a'.

It remains to prove Claim 7. We shall give two different proofs.

First proof of Claim 7: For $n \in \omega$, define recursively sets c_n as follows:

$$\begin{aligned} c_0 &= f[a \backslash a']; \\ c_{n+1} &= f[c_n]. \end{aligned}$$

Let $c^* = \bigcup_{n \in \omega} c_n$. Then $c^* \subseteq rng(f) \subseteq a''$. We show that $f[c^* \cup (a \backslash a')] = c^*$. Let $x \in c^* \cup (a \backslash a')$. If $x \in a \backslash a'$, then $f(x) \in c_0 \subseteq c^*$. If $x \in c^*$, then $x \in c_n$ for some $n \in \omega$, and hence, $f(x) \in c_{n+1} \subseteq c^*$. In either case, $x \in c^* \cup (a \backslash a')$ implies $f(x) \in c^*$; and we have shown the inclusion $f[c^* \cup (a \backslash a')] \subseteq c^*$. For the other inclusion, assume $y \in c^*$. If $y \in c_0$, then there exists an $x \in a \backslash a'$ such that $f(x) = y$. If $y \in c_{n+1}$ for some $n \in \omega$, then there exists an $x \in c_n$ such that $f(x) = y$. In either case, $y \in f[c^* \cup (a \backslash a')]$, and we have proved Claim 7.

Second proof of Claim 7: Let

$$C = \{c \subseteq a'' : f[c \cup (a \setminus a')] \subseteq c\}.$$

We call the sets in C *candidates* (for c^*). Clearly, $a'' \in C$. For $c \in C$ we denote $c' = f[c \cup (a \setminus a')]$. Then

$$f[c' \cup (a \setminus a')] = f[f[c \cup (a \setminus a')] \cup (a \setminus a')] \subseteq f[c \cup (a \setminus a')] = c',$$

and hence,

(1) $$\forall c \in C\, (f[c \cup (a \setminus a')] \in C).$$

Let $c^* = \bigcap C$. Then

$$f[c^* \cup (a \setminus a')] = f\left[\bigcap_{c \in C} (c \cup (a \setminus a'))\right] \subseteq \bigcap_{c \in C} f[c \cup (a \setminus a')] \subseteq \bigcap_{c \in C} c = c^*,$$

and it follows that c^* is the smallest candidate.

It remains to show that $f[c^* \cup (a \setminus a')] = c^*$. On the one hand, since c^* is a candidate, we have $f[c^* \cup (a \setminus a')] \subseteq c^*$. On the other hand, by (1), the set $f[c^* \cup (a \setminus a')]$ is also a candidate, and therefore $c^* \subseteq f[c^* \cup (a \setminus a')]$. This concludes the proof of Claim 7. □

What happened in the second proof to our recursive definition? It was replaced by taking the intersection of a family of sets.

EXERCISE 7(PG): Show that the sets c^* constructed in the two proofs of Claim 7 coincide.

As briefly indicated in Chapter 3, a set a is *finite* if $a \approx n$ for some natural number n. The question arises whether one can characterize the finite sets without explicit reference to natural numbers. Now we are going to explore a possible way of doing this.

Theorem 8: *Let $n = \{0, 1, \ldots, n-1\}$ be a natural number, and let $f : n \longrightarrow n$ be a one-to-one function. Then f maps n onto n.*

Proof: By induction over n. For $n = 0$, the only function $f : 0 \longrightarrow 0$ is the empty set. The range of this function is $\emptyset = 0$; and hence f is a surjection.

Assume now that every one-to-one function $f : n \longrightarrow n$ maps n onto n, and let $g : n+1 \longrightarrow n+1$. Suppose toward a contradiction that g is one-to-one, but not onto.

Let $k \in n+1 \backslash rng(g)$. Define a function $p : n+1 \to n+1$ by: $p(k) = n$, $p(n) = k$, and $p(m) = m$ for $m \notin \{k, n\}$. Then p is a permutation[2] of $n+1$, and thus the composition $h = p \circ g$ is a one-to-one function from $n+1$ into $n+1$. It is not hard to see that $rng(h) \subseteq n$. Therefore, the restriction $h|n$ is a one-to-one function from n into n. By our inductive assumption, $h|n$ maps n onto n. In particular, since $h(n) \in n$, there is some $m < n$ such that $h(m) = h(n)$. Thus h is not one-to-one. We have reached a contradiction. □

EXERCISE 8(G): Identify the three parts (I1), (I2) and (I3) in the proof of Theorem 8.

Theorem 9: *Let $n = \{0, 1, \ldots, n-1\}$ be a natural number, and let $f : n \to n$ be a surjection. Then f is also an injection.*

EXERCISE 9(G): Prove Theorem 9.

Corollary 10: *Let a be a finite set, and let $f : a \to a$. Then f is an injection iff it is a surjection.*

EXERCISE 10(G): Prove Corollary 10.

EXERCISE 11(G): Prove that ω is infinite.

Theorem 11: *Let A be an infinite set. Then there exists a one-to-one function $f : \omega \to A$.*

Proof: We construct recursively the function values $f(n)$. A is infinite; in particular, A is nonempty. We let $f(0)$ be any element of A.

Suppose $f|n : n \to A$ has already been constructed, and is an injection. Since A is infinite, $f|n$ is not a surjection, and we can choose some $a_n \in A \backslash rng(f|n)$ and set $f(n) = a_n$. The resulting function $f|(n+1)$ will also be an injection.

Let us show that the function f is one-to-one: Suppose toward a contradiction that $f(n) = f(m)$ for $n < m$. This is impossible, since $f(m)$ was chosen from $A \backslash rng(f|m)$, and $f(n) \in rng(f|m)$. □

EXERCISE 12(F): Identify parts (R1), (R2) and (R3) of the recursive construction in the proof of Theorem 11. Which part of the proof does not belong to the recursive construction? Part (R2) contains an inductive assumption.[3] What is it?

Wait a minute. This doesn't really look like a recursive definition. In Definition 2, which is our prototype of a recursive definition, we specified unequivocally for any given X^k what X^{k+1} was supposed to be. Here, if an injection $f|n$ is given, we argue that one *can* find $f(n)$ so that $f|n+1$ will also be one-to-one, but we do not say clearly which element of $A \backslash rng(f|n)$ should be chosen.

This is an important point. Note that we called the procedure used in this proof a *recursive construction* rather than a recursive definition. We shall make

[2] One-to-one functions from a set x onto itself are often called *permutations of x*.

[3] Even those who meticulously distinguish between induction and recursion say "inductive" in this case.

this distinction when we want to indicate whether all steps in a recursion contain explicit prescriptions of which element to choose next (a recursive definition), or merely arguments that the next step in the recursion can be carried out (a recursive construction). Most contemporary set theorists feel comfortable with recursive constructions and use them a lot. Nevertheless, one should replace recursive constructions by recursive definitions whenever possible. In the proof of Theorem 11, this is impossible though. We shall discuss this matter in more depth in Chapter 9.2.

EXERCISE 13(PG): Do you still have a copy of your proof of Theorem 2 of Chapter 2? If so, check whether by any chance you used an inductive definition or an inductive construction in this proof. If you did not succeed in proving Theorem 2.2, try again. We just gave you a hint.

Now we present some interesting consequences of Theorem 11.

Corollary 12: *Let A be a subset of ω. Then A is either finite or denumerable.*

Proof: If A is a subset of ω, then $|A| \leq |\omega| = \aleph_0$. Theorem 11 implies that if A is infinite, then $|A| \geq \aleph_0$. Now the Cantor-Schröder-Bernstein Theorem implies that A is either finite or of cardinality $\aleph_0$. □

EXERCISE 14(PG): Give a direct proof of Corollary 12 that uses a recursive definition instead of a reference to Theorem 11, which was proved by means of a recursive construction.

Corollary 13: *If A is an infinite set, then there exists a function $f : A \to A$ that is one-to-one, but not onto.*

Proof: Suppose A is infinite. By Theorem 11, there exists an injection $h : \omega \to A$. Fix such an h, and define a function $f : A \to A$ by:

$$f(x) = \begin{cases} h(n+1) & \text{if } x = h(n) \text{ for some } n \in \omega; \\ x & \text{otherwise.} \end{cases}$$

EXERCISE 15(G): Show that the function f is one-to-one but not onto. □

Corollary 14: *If A is an infinite set, then there exists a function $f : A \to A$ that is onto, but not one-to-one.*

EXERCISE 16(G): Prove Corollary 14.

Now we can characterize the finite sets without reference to natural numbers.

Theorem 15: *For any set A, the following conditions are equivalent:*

(a) *A is finite.*
(b) *Every injection $f : A \to A$ is a surjection.*
(c) *Every surjection $f : A \to A$ is an injection.*

In Chapter 2, we promised to eventually show how the order types $\eta = ot(\langle \mathbb{Q}, \leq \rangle)$ and $\lambda = ot(\langle \mathbb{R}, \leq \rangle)$ can be characterized. The following theorem is due to Georg

Cantor. Its proof involves another interesting example of a recursive construction.

Theorem 16: *Let $\langle A, \preceq\rangle$ be a dense l.o. without minimum and maximum elements. If $|A| = \aleph_0$, then $\langle A, \preceq\rangle \cong \langle \mathbb{Q}, \leq\rangle$.*

EXERCISE 17(G): (a) Conclude from Theorem 16 that

$$\langle \mathbb{Q}, \leq\rangle \cong \langle \mathbb{Q} \cap ((0,1) \cup [2,3)), \leq\rangle.$$

(b) Compare the effort it took to get the solution of (a) with the amount of work spent to do Exercise 23 of Chapter 2.

Proofs involving recursive constructions may contain a large number of symbols, which often makes them difficult to read. The best policy is to read such proofs backwards.[4] More often than not, the backward reading will correspond to the order of steps which the author of the proof took in discovering it. This way, you will also learn how to find such proofs yourself. As a demonstration of this backward reading technique, we first give you a standard proof of Theorem 16. Next, we take you on a backward tour through this proof. For your convenience, the proof itself is printed in italics, so that you can easily skip ahead to the starting point of our tour.

Proof of Theorem 16: (P1) *Let $\langle A, \preceq\rangle$ be as in the assumptions. Recall that $\langle \mathbb{Q}, \leq\rangle$ denotes the set of rationals with the usual order. Extend $\langle A, \preceq\rangle$ and $\langle \mathbb{Q}, \leq\rangle$ to $\langle \overline{A}, \preceq\rangle$ and $\langle \overline{\mathbb{Q}}, \leq\rangle$ by adding largest elements: a_∞ to A and ∞ to $\mathbb{Q}$, as well as smallest elements: $a_{-\infty}$ to A and $-\infty$ to $\mathbb{Q}$. Since $\langle A, \preceq\rangle$ and $\langle \mathbb{Q}, \leq\rangle$ are dense l.o.'s without endpoints, the extensions $\langle \overline{A}, \preceq\rangle$ and $\langle \overline{\mathbb{Q}}, \leq\rangle$ are also dense. Let $\langle a_i : i \in \omega\rangle$ be an enumeration of A, and let $\langle r_i : i \in \omega\rangle$ be an enumeration of $\mathbb{Q}$. We construct a sequence $\langle f_i : i \in \omega\rangle$ of finite functions such that for $i < k$:*

(i) *$f_i \subseteq f_k$;*

(ii) *f_i is order-preserving;*

(iii) *$\{a_j : j < i\} \subseteq dom(f_i) \subseteq \overline{A}$;*

(iv) *$\{r_j : j < i\} \subseteq rng(f_i) \subseteq \overline{\mathbb{Q}}$.*

(P2) *We let $f_0 = \{\langle a_{-\infty}, -\infty\rangle, \langle a_\infty, \infty\rangle\}$. Suppose f_i has already been constructed. We first construct a function $g_i \supseteq f_i$. If $a_i \in dom(f_i)$, then $g_i = f_i$. Now suppose $a_i \notin dom(f_i)$. Let $dom(f_i) = \{d_j : j \leq k\}$, where $d_0 < d_1 < \cdots < d_k$. For $j < k$, denote $I_j = (d_j, d_{j+1})$. Let j_0 be such that $a_i \in I_{j_0}$.*

We choose $r \in (f(d_{j_0}), f(d_{j_0+1}))$ and let $g_i = f_i \cup \{\langle a_i, r\rangle\}$.

If $r_i \in rng(g_i)$, then we define $f_{i+1} = g_i$.

Otherwise, let $rng(g_i) = \{e_j : j < \ell\}$, where $e_0 < e_1 < \cdots < e_\ell$. For $j < \ell$, let $J_j = (e_j, e_{j+1})$, and let j_1 be such that $r_i \in J_{j_1}$. We choose $a \in (g_i^{-1}(e_{j_1}), g_i^{-1}(e_{j_1+1}))$, and let $f_{i+1} = g_i \cup \{\langle a, r_i\rangle\}$. Obviously, f_{i+1} satisfies conditions (i)–(iv).

(P3) *Then $f = \bigcup_{i\in\omega} f_i$ is an order isomorphism between $\langle \overline{A}, \preceq\rangle$ and $\langle \overline{\mathbb{Q}}, \leq\rangle$. It follows that the restriction $f|A$ is an order isomorphism between $\langle A, \preceq\rangle$ and $\langle \mathbb{Q}, \leq\rangle$.*

This proof is a rather typical example of how proofs by recursive construction are written in research papers. Now we are going to show you how you can speed up the reading of such a proof, and how you can find such proofs yourself.

---

[4]This should not be taken too literally. We'll show you in a moment what we mean.

"Let $\langle A, \preceq\rangle$ be as in the assumptions." You should always start like this, no matter whether you are writing down a proof, or whether you are trying to discover one. In advanced publications, these opening statements are often omitted in order to save space and time. However, even experienced theorem provers like to start the search for a proof by jotting down such a sentence. It helps in getting one's mind focused.

The next step in discovering a proof should be the question: "What do we want to accomplish?" In the case of Theorem 16, we want to find an order isomorphism $f : A \longrightarrow \mathbb{Q}$. Our next question should be: "How should we go about the business of finding such a function f?" At this stage of the search, a very general answer is called for. We have basically three options: We could try to find an explicit formula for f, we could try to infer f's existence from some general existence theorem (without actually constructing f), or we could try to use a recursive construction. Beginners often become too much preoccupied with details at this stage of the search. This is counterproductive, since exploring one of the options in detail may lead into a blind alley. If this happens, the best thing is to return to the initial list of options and try a different approach. You will be able to do so only if you keep an initial list of options handy, be it in the back of your mind or on paper.

Once we have decided what our basic options are, we can determine which of them looks most promising. In our case, trying to find an explicit formula for f is hopeless, since we know very little about A. Using a general existence theorem might be a better bet. Let's see whether we know such a theorem. Theorem 16 would be just right, but we cannot use it, since this is the theorem we are about to prove. Too bad. So, perhaps we should explore the third avenue and look for a recursive construction of f.

Experienced mathematicians perform the listing and evaluation of the available options so quickly and routinely that they often are not even aware of what they are doing. For this reason, authors of proofs do not always tell you explicitly what kind of construction they use. The author of a proof may expect that you, the reader, will have figured out on your own what kind of proof you are going to see.[5] Neither of the words "inductive(ly)" or "recursive(ly)" appears anywhere in the proof of Theorem 16.

Now let us see how the recursive construction has been done in the proof of Theorem 16. Let us begin by looking at the last part (P3) of this proof. We see that a function f is being defined as a union of f_i's. Apparently, the f_i's have been constructed by recursion earlier in the proof. A look at (P3) also reveals two other features of the proof: For whatever reason, the authors of the proof introduced p.o.'s $\langle \overline{A}, \preceq\rangle$ and $\langle \overline{\mathbb{Q}}, \leq\rangle$ that extend the original l.o.'s $\langle A, \preceq\rangle$ and $\langle \mathbb{Q}, \leq\rangle$, and constructed f as an isomorphism of these extended p.o.'s.

EXERCISE 18(G): (a) How can we tell from (P3) alone that $\langle \overline{A}, \preceq\rangle$ and $\langle \overline{\mathbb{Q}}, \leq\rangle$ are extensions of $\langle A, \preceq\rangle$ and $\langle \mathbb{Q}, \leq\rangle$?

(b) Could we also conclude from (P3) alone that $\langle \overline{A}, \preceq\rangle$ and $\langle \overline{\mathbb{Q}}, \leq\rangle$ are l.o.'s rather than just p.o.'s?

[5]We do not want to encourage you to write like this yourself. Whenever you are going to present a recursive construction, say so. But you want to learn how to read real research papers, including poorly written ones.

Moreover, the authors of the proof do not bother showing explicitly at the end of their proof that f is an order isomorphism. This is a common practice. It should be used only if it is evident from the inductive assumptions made at each stage of the construction that the object being constructed has the desired properties.

What kind of propeties of the f_i's would guarantee that f is as desired? First of all, f needs to be a function. Therefore, each f_i should also be a function; more precisely, each f_i should be a function whose domain is contained in $\overline{A}$ and whose range is contained in $\overline{\mathbb{Q}}$ (whatever these sets may be). Moreover, if $a \in dom(f_i) \cap dom(f_j)$, then we must be sure that $f_i(a) = f_j(a)$. There are two convenient ways of achieving the latter. One possibility is to require that the domains of the f_i's are pairwise disjoint. The other option is to require that $f_i \subseteq f_{i+1}$ for all $i \in \omega$. In the proof we are analyzing here, the second option was used.

EXERCISE 19(G): Show that if $h = \bigcup_{n\in\omega} h_n$, each h_n is a function, and $h_n \subseteq h_{n+1}$ for all $n \in \omega$, then h is also a function.

Furthermore, f must be order-preserving. This requirement is similar to the requirement that f is a function.

EXERCISE 20(G): Let $\langle X, \preceq_X\rangle$ and $\langle Y, \preceq_Y\rangle$ be p.o.'s. Suppose that for each $n \in \omega$, an order-preserving function h_n from a subset of X into Y is given. Show that if $h_n \subseteq h_{n+1}$ for each $n \in \omega$, then the function $h = \bigcup_{n\in\omega} h_n$ is order-preserving.

Therefore, if we require that for each $i \in \omega$ the inclusion $f_i \subseteq f_{i+1}$ holds and f_i is an order preserving function from a subset of $\overline{A}$ into $\overline{\mathbb{Q}}$, then f will be an order-preserving function from a subset of $\overline{A}$ onto a subset of $\overline{\mathbb{Q}}$. By Exercise 2.19, f will also be an injection. But we want the domain of f to be *all* of $\overline{A}$, and its range to be *all* of $\overline{\mathbb{Q}}$. How can we assure this? The trick is to enumerate $\overline{A}$ and $\overline{\mathbb{Q}}$ by natural numbers, and then to require that the i–th element of the enumeration of $\overline{A}$ is in the domain of f_{i+1}, and the i–th element of the enumeration of $\overline{\mathbb{Q}}$ is in the range of f_{i+1}.

Now we are ready to read the first part (P1) of the proof. The intentions of its authors should be clear by now, except that we still don't know why they saw fit to add first and last elements to the l.o.'s. It is quite common that early parts of proofs contain definitions whose motivation will emerge only later in the proof. Sometimes it is possible to second-guess the author's intention, and you should always give this a try. But don't try too hard. It is often best to read on and discover the motivation for a concept when you see how this concept is being used. Therefore, let us now turn to (P2).

EXERCISE 21(G): Or have you not even gone over (P1) yet? If so, do it now.

The first line of (P2) just describes how to get started. This sentence corresponds to part (R1) of Definition 2. The rest of (P2) describes how, given a function f_i that satisfies (ii)–(iv), we can[6] find a function f_{i+1} such that f_{i+1} satisfies conditions (i)–(iv). In extending f_i to f_{i+1}, one must obey several rules, and this will be easier if one performs the construction in two stages. At the first of these two

[6] Remember: this is a recursive construction, not a recursive definition.

stages, the authors take care of condition (iii). This condition requires a_i to be in $dom(f_{i+1})$. If $a_i \notin dom(f_i)$, then they will have to assign a function value to a_i. In doing this, they must take care not to violate condition (ii). So they consider the smallest interval (a_{j_0}, a_{j_0+1}) of $\overline{A}$ with endpoints in $dom(f_i)$ that contains a_i. Then, they pick some $r \in (f_i(a_{j_0}), f_i(a_{j_0+1}))$ and let $g_i(a_i) = r$. This choice of $g_i(a_i)$ guarantees that g_i is order-preserving as long as f_i is. Note that the authors of the proof do not tell you explicitly why such an r exists. They leave it to you, the reader, to figure that out. We do not recommend that you start leaving little gaps in your own proofs. But if you want to read research papers written by other people, then you have to learn how to fill in some of the gaps in the arguments yourself. Here is a useful technique for doing this: Whenever you read a proof, carefully check where all the assumptions of the theorem are used. Do this not only for the explicit assumptions of the theorem, but also for hidden ones. For example, the question "Where was denseness of $\langle \mathbb{Q}, \leq \rangle$ used in this proof?" leads immediately to the reason why a suitable candidate r for $g_i(a_i)$ always exists. Note that denseness of $\langle \mathbb{Q}, \leq \rangle$ is not mentioned in the wording of the theorem.

EXERCISE 22(PG): How do we know that denseness of $\langle \mathbb{Q}, \leq \rangle$ has to be used at *some* point in *any* proof of Theorem 16?

Okay, we have had enough complaints about the authors of this proof (as you may have suspected all along, the proof was written by ourselves). Let us say something nice for a change: It turns out that passing to the extensions $\langle \overline{A}, \preceq \rangle$ and $\langle \overline{\mathbb{Q}}, \leq \rangle$ of the original orders was a clever device for streamlining the write-up of part (P2). Were it not for these extensions, the case where a_i is smaller (or larger than) all the elements in $dom(f_i)$ would have to be considered separately.

The kind of construction used in this proof is called a *back and forth construction* of a function. The procedure can be understood as alternating between the construction of little pieces of f ("forth") and little pieces of f^{-1} ("back"). The step from f_i to g_i is the "forth" direction of the construction. It takes care of requirement (iii). In order to take care of requirement (iv), one must go "back" and construct $f^{-1}(r_i)$. This is is being done in the second half of (P2) by extending g_i to f_{i+1}.

The last sentence of (P2) contains the word "obviously." This is one of the many synonyms for our favored word EXERCISE.

EXERCISE 23(G): Convince yourself that the function f_{i+1} indeed meets all the requirements it is supposed to meet.

EXERCISE 24(G): Where in the proof of Theorem 16 was the assumption of denseness of $\langle A, \preceq \rangle$ used? How about the assumption that this l.o. has no endpoints? How about countability of A?

We conclude this section with a characterization of the order type λ of $\langle \mathbb{R}, \leq \rangle$.

Let $\langle A, \preceq \rangle$ be a l.o., and let $X, Y \subseteq A$. We write $X \prec Y$ if $\forall x \in X \forall y \in Y\, (x \prec y)$. If $a \in A$, then we write $X \prec a$ instead of $X \prec \{a\}$. The symbols $X \preceq Y$, $X \preceq a$ and $a \preceq Y$ are defined analogously. A pair $\langle X, Y \rangle$ such that $X, Y \neq \emptyset$, $X \cup Y = A$, and $X \prec Y$ is called a *Dedekind cut*. A Dedekind cut X, Y is called a *gap* if there is no $a \in A$ with $X \preceq a$ and $a \preceq Y$.

EXERCISE 25(G): Convince yourself that if $\langle A, \preceq\rangle = \langle \mathbb{Q}, \leq\rangle$, then the pair $\langle\{q \in \mathbb{Q} : q < \sqrt{2}\}, \{q \in \mathbb{Q} : q > \sqrt{2}\}\rangle$ is a gap.

EXERCISE 26(G): Suppose that $\langle X, Y\rangle$ is a Dedekind cut in a l.o. $\langle A, \preceq\rangle$. Show that if $\langle X, Y\rangle$ is not a gap, then there exists a *unique* $a \in A$ such that $X \preceq a \preceq Y$.

A l.o. $\langle A, \preceq\rangle$ is *complete* (or *Dedekind complete*) if it has no gaps. The l.o. $\langle \mathbb{R}, \leq\rangle$ is an example of a complete l.o.

EXERCISE 27(G): Convince yourself that Dedekind completeness is an order invariant; i.e., show that if $\langle X \preceq_X\rangle$ and $\langle Y, \preceq_Y\rangle$ are isomorphic l.o.'s, then $\langle X, \preceq_X\rangle$ is Dedekind complete iff $\langle Y, \preceq_Y\rangle$ is.

If $\langle A, \preceq\rangle$ is a dense linear order without minimum and maximum elements, and $D \subseteq A$, then we say that D is *dense in* $\langle A, \preceq\rangle$ if

$$\forall a, b \in A\,(a \prec b \rightarrow \exists d \in D\,(a \prec d \wedge d \prec b)).$$

Lemma 17: *Let $\langle A_0, \preceq_0\rangle$, $\langle A_1, \preceq_1\rangle$ be dense, complete linear orders without largest or smallest elements, and let D_0, D_1 be dense subsets of A_0 and A_1 respectively. For every order-preserving function $f : D_0 \rightarrow D_1$ there exists an order-preserving function $\tilde{f} : A_0 \rightarrow A_1$ such that $\tilde{f}|D_0 = f$. Moreover, if f maps D_0 onto D_1, then $\tilde{f}$ maps A_0 onto A_1.*

Proof: Let A_0, A_1, D_0, D_1 be as in the assumptions, and let $f : D_0 \rightarrow D_1$ be order-preserving. For $r \in A_0$ denote $X_r = \{b \in A_1 : \exists d \in D_0\,(d \preceq_0 r \wedge b \preceq_1 f(d))\}$, and $Y_r = A_1 \backslash X_r$.

EXERCISE 28(G): Convince yourself that the pair $\langle X_r, Y_r\rangle$ is a Dedekind cut in $\langle A_1, \preceq_1\rangle$.

Now define: $\tilde{f}(r) = a(X_r, Y_r)$, where $a(X_r, Y_r)$ is the unique $a \in A_1$ with $X_r \preceq_1 a \preceq_1 Y_r$.

EXERCISE 29(PG): Prove that the function $\tilde{f}$ is as required. □

As we already mentioned, the l.o. $\langle \mathbb{R}, \leq\rangle$ is complete, and it contains no smallest and no largest element. Moreover, the set $\mathbb{Q}$ of rationals is a countable dense subset of $\mathbb{R}$. Combining Theorem 16 and Lemma 17, we obtain the following characterization of $ot(\langle \mathbb{R}, \leq\rangle)$.

Theorem 18: *Every complete, dense l.o. without endpoints which has a countable dense subset is order-isomorphic to $\langle \mathbb{R}, \leq\rangle$.*

EXERCISE 30(G): Derive Theorem 18 from Theorem 16 and Lemma 17.

4.2. Induction and recursion over wellfounded sets

Now we shall take a more inquisitive look at the last part (I3) of an inductive proof. Let P be a property of natural numbers. Instead of "n has property P," we shall write "$P(n)$." Consider the following:

Schematic Theorem: *$P(n)$ holds for all $n \geq n_0$.*

Many theorems that have inductive proofs are of this form. For example, in Theorem 1, $n_0 = 1$ and n has property P if $|\mathbb{N}^n| = \aleph_0$.

EXERCISE 31(G): Find n_0 and formulate a property P such that Theorem 8 becomes a special case of the Schematic Theorem.

EXERCISE 32(G): Outline parts (I1) and (I2) of an inductive proof of the Schematic Theorem.

The last part of an inductive proof of the Schematic Theorem says:

(I3) *Since $P(n_0)$, and $\forall k \geq n_0\,(P(k) \rightarrow P(k+1))$, all natural numbers $n \geq n_0$ have property P.*

Is that so? Why? Suppose it were not true that $P(n)$ holds for all natural numbers $n \geq n_0$. Then the set of counterexamples $C = \{n \in \omega : n \geq n_0 \wedge P(n)$ does not hold$\}$ would be nonempty. Since $\langle \omega\backslash n_0, \leq \rangle$ is a w.o., the set C would have a minimum element n_1. Consider two cases.

Case 1: $n_1 = n_0$. This is impossible by part (I1) of the inductive proof.

Case 2: $n_1 \neq n_0$. Since C consists by definition only of numbers $\geq n_0$, we must have $n_1 > n_0$. Then $n_1 = n_2 + 1$ for some $n_2 \geq n_0$. Moreover, $n_2 \notin C$, since n_1 was assumed to be the minimum element of C. Therefore, $P(n_2)$ holds. By (I2), the implication $P(n_2) \rightarrow P(n_2 + 1)$ holds. Hence n_1 has property P, and we have reached a contradiction.

In other words, the combination of (I1) and (I2) constitutes a proof of $P(n)$ for all natural numbers $n \geq n_0$, because the set of these numbers is wellordered, and every $n_1 > n_0$ is the immediate successor of some $n_2 \geq n_0$.

The question arises whether we need all these properties of $\langle \omega\backslash n_0, \leq \rangle$ for inductive arguments.

Consider the example of the following property $P(n)$: "There exists a prime number p such that $p|n$."

Fact 19: *$P(n)$ holds for every natural number $n \geq 2$.*

First proof of Fact 19: By induction. Let $n \geq 2$. Suppose $P(k)$ holds for all natural numbers k such that $n > k \geq 2$. We show that $P(n)$ holds. Consider two cases.

Case 1: n is prime. Then $P(n)$ holds since $n|n$.

Case 2: n is composite. Then $n = k \cdot m$, where $2 \leq k \leq m < n$. By assumption, $P(k)$ holds, i.e., there exists some prime p such that $p|k$. Fix such a p, and note that $p|n$. This proves $P(n)$.

Is this a proof by induction? It looks like one, but does it have distinguishable parts (I1)–(I3)? Evidently, part (I3) has been condensed to the phrase "By induction." Let us be a little more explicit about this part of the argument. Again, suppose toward a contradiction that there is some $n \in \omega\backslash 2$ such that $P(n)$ does not hold. Then the set $C = \{n \in \omega\backslash 2 : P(n) \text{ does not hold}\}$ is nonempty, and hence it has a minimal element n_1. Since n_1 is a minimal counterexample, $P(n)$ holds for all $2 \leq n < n_1$. The proof of Fact 19 shows that $P(n_1)$ also holds, which contradicts the choice of n_1.

This is even shorter than the justification of (I3) for a standard inductive proof. How come? Why, for example, did we not have to give special consideration to the case $n = 2$? Or did we? After all, the above proof of Fact 19 deals with two separate cases. But why does the first case concern all primes, rather than just $n = 2$? In order to answer these questions, let us slightly modify the proof of Fact 19.

Second proof of Fact 19: Let $n \in \omega\backslash 2$. Suppose $P(k)$ holds for all $k \in \omega\backslash 2$ such that $k|n$ and $k \neq n$. We show that $P(n)$ holds. Consider two cases.

Case 1: n is prime. Then $P(n)$ holds since $n|n$.

Case 2: n is composite. Then $n = k \cdot m$, where $k|n$ and $k \neq n$. By assumption, $P(k)$ holds, i.e., there exists some prime p such that $p|k$. Fix such a p, and note that $p|n$. This proves $P(n)$.

This argument is even an ε shorter than the first proof.

EXERCISE 33(G): Write an explanation why this is a proof of Fact 19. Model your explanation on our discussion of the first proof.

Congratulations! You have discovered the idea of induction over a wellfounded set. In order to express your discovery in the form of an abstract theorem, let us generalize some concepts from the theory of partial order relations to arbitrary wellfounded relations.

Definition 20: Let X be a nonempty set, let $x \in X$, and $Y \subseteq X$. Assume that W is a wellfounded relation on X. The set $\{z \in X : \langle z, x\rangle \in W \wedge z \neq x\}$ is called the *initial segment of* $\langle X, W\rangle$ *determined by* x, and denoted by $I_W(x)$ (or simply $I(x)$ if the relation W is implied by context). We say that x *is* W*-minimal for* Y if $x \in Y$ and $I(x) \cap Y = \emptyset$. If x is W-minimal for X, then we simply say that x *is* W*-minimal*.

Theorem 21 (Principle of Induction over a Wellfounded Set): *Let X be a nonempty set, let $Y \subseteq X$, and let W be a wellfounded relation on X. If the implication*

$$(\mathrm{Ind}_W(x)) \qquad I_W(x) \subseteq Y \rightarrow x \in Y$$

holds for all $x \in X$, then $X = Y$.

EXERCISE 34(G): Let $X = \omega$, $Y = \{x : 2|x\}$, and let W be the natural order $\leq$ on ω.

(a) Show that ($\mathrm{Ind}_W(13)$) holds.

(b) Conclude that Theorem 21 cannot be strengthened to the assertion that for all $x \in X$, if ($\mathrm{Ind}_W(x)$) holds, then $x \in Y$.

(c) Visualize the difference between the assertion of point (b) and Theorem 21 by writing the two statements with quantifiers, brackets and logical connectives. Convince yourself that the example given in point (a) does not contradict Theorem 21.

Proof of Theorem 21: Let X, Y, and W be as in the assumption, and assume that ($\mathrm{Ind}_W(x)$) holds for all $x \in X$. Suppose toward a contradiction that the set $C = X \backslash Y$ is nonempty. By wellfoundedness of W, there exists a W-minimal element of C. But, if x_0 is a W-minimal element of C, then $I_W(x_0) \subseteq Y$, and it follows from ($\mathrm{Ind}_W(x_0)$) that $x_0 \in Y$, which is impossible, since $x_0 \in C = X \backslash Y$. □

Sounds familiar? Your solution to Exercise 34 is just the special case of the proof of Theorem 21 for the wellfounded p.o. $\langle \omega \backslash 2, | \rangle$. In other words, in the second proof of Fact 19, we establish that the implication ($\mathrm{Ind}_W(n)$) holds for every $n \in \omega \backslash 2$, where the role of W is played by the wellfounded partial order relation $|$, and Y is the set $\{n \in \omega \backslash 2 : P(n)\}$. In Case 1 of this proof, we establish the implication ($\mathrm{Ind}_W(n)$) for $|$-minimal n's. In Case 2, we do the same for the remaining n's.

There are several ways of adding the missing part (I3) to the second proof of Fact 19. Since we have explicitly formulated Theorem 21, it is possible to rewrite the whole proof in the terminology of this theorem, and conclude it with the sentence: "By the Principle of Induction over Wellfounded Sets, $P(n)$ holds for every natural number $n \geq 2$."

EXERCISE 35(G): Rewrite the second proof of Fact 19 in this style.

One could also start with the phrase: "By induction over the wellfounded p.o. $\langle \omega \backslash 2, | \rangle$," and leave the remainder of the second proof of Fact 19 unchanged. The latter style is less formal than that of your solution to Exercise 35, but it is acceptable for most types of mathematical writing.

How is the first proof of Fact 19 related to the second one? In essence, the two proofs are the same. As we mentioned in Chapter 2, the wellorder relation $\leq$ extends the relation $|$ on $\omega \backslash 2$. Therefore, it is possible to write the proof of Fact 19 as a standard argument by induction over $\langle \omega \backslash 2, \leq \rangle$. This is what we did in the first proof. However, we do not recommend this style of writing. It only makes the proof longer and obscures its structure.

The next theorem is an important application of Theorem 21.

Theorem 22: *Let $\langle X, \preceq \rangle$ be a w.o., and let $f : X \to X$ be order-preserving. Then $x \preceq f(x)$ for all $x \in X$.*

Proof: Let $\langle X, \preceq \rangle$ and f be as in the assumptions. Denote $Y = \{x \in X : x \preceq f(x)\}$. We want to show that $Y = X$. Since wellorder relations are wellfounded, it suffices to show that the implication

$$(\mathrm{Ind}_W(x)) \qquad (\forall y \prec x \, (y \preceq f(y))) \rightarrow x \preceq f(x)$$

holds for all $x \in X$. Suppose towards a contradiction that $x \in X$ is such that $(\text{Ind}_W(x))$ fails. Then $f(x) = y \prec x$, where y is such that $y \preceq f(y)$. This is impossible, since f was assumed to be order-preserving. □

EXERCISE 36(G): Where in the above proof did we use the result of Exercise 2.19?

Is there also such a thing as recursion over wellfounded sets? Yes. Let us try to discover this technique by taking a close look at how sequences $\langle x_n : n \in \omega\rangle$ are constructed by recursion over the set of natural numbers. A sequence $\langle x_n : n \in \omega\rangle$ is just a function $F : \omega \to X$, where X is some set, and $F(n) = x_n$ for all $n \in \omega$. One starts a recursion by saying what $F(0)$ is, and then one proceeds by specifying the possible choices of $F(n)$ for given $F(0), \dots , F(n - 1)$. In the examples we have seen so far, $F(n)$ depends only on $F(n - 1)$. This need not be the case in general. For instance, the *Fibonacci sequence* $\langle Fi(n) : n \in \omega\rangle$ is defined recursively by $Fi(0) = Fi(1) = 1$, and $Fi(n) = Fi(n - 2) + Fi(n - 1)$ for $n > 1$.

Recursion comes in two varieties: recursive definitions and recursive constructions. In a recursive definition, we are given a set X, the function value $F(0)$ (which is an element of X), and a function G that is defined at least for all finite sequences of the form $\langle F(0), \dots , F(n - 1)\rangle$ and takes values in X. Then we define $F(n) = G(F(0), \dots , F(n - 1))$ for all $n > 0$. The very nature of recursion requires that the function G should be given *before* the construction of F. Therefore, it is too restrictive to demand that the domain of G consists only of sequences of the form $\langle F(0), \dots , F(n - 1)\rangle$. The most flexible approach is to assume that the domain of G consists of *all* finite sequences of elements of X. The set of all these sequences will be denoted by ${}^{<\omega}X$. Thus, G is a function from ${}^{<\omega}X$ into X. Most of the domain of G will not be used in the construction of F, but this does not hurt. Since the empty set can be treated as a finite sequence (with empty domain) of elements of X, we can require that $F(0) = G(\emptyset)$, which eliminates the need for a separate definition of $F(0)$.

For example, if $G(\emptyset) = \emptyset$ and $G(F(0), \dots , F(n - 1)) = F(n - 1) \cup \{F(n - 1)\}$ for all $n > 0$, then we obtain the function $F(n) = \widehat{n}$ of Definition 3′.

EXERCISE 37(G): Which function could serve as G in Definition 4(a)? *Hint:* Don't worry too much about the choice of an appropriate set X. Just assume that there is a set X which contains all relevant objects. In Chapter 7 we shall explain why such an approach is permissable.

Let us now explicitly state the theorems that we have been using all along.

Theorem 23 (Principle of Recursive Definitions): *Suppose X is a nonempty set, and let $G : {}^{<\omega}X \to X$ be given. Then there exists exactly one function $F : \omega \to X$ such that $F(0) = (\emptyset)$ and $F(n) = G(F(0), \dots , F(n - 1))$ for all $n > 0$.*

Theorem 24 (Principle of Recursive Constructions): *Suppose X is a nonempty set, and let $G^* : {}^{<\omega}X \to \mathcal{P}(X)\backslash\{\emptyset\}$ be given. Then there exists a function $F : \omega \to X$ such that $F(0) \in G^*(\emptyset)$ and $F(n) \in G^*(F(0), \dots , F(n - 1))$ for all $n > 0$.*

EXERCISE 38(G): All one really does in a recursive construction is check that at each stage of the construction a suitable $F(n)$ exists. This amounts to verifying one of the assumptions of Theorem 24. Which assumption are we talking about?

EXERCISE 39(G): Find a function that could serve as G^* in the recursive construction involved in the proof of Theorem 11.

Now we are going to generalize Theorems 23 and 24.

Theorem 25 (Principle of Generalized Recursive Definitions): *Let $X, Z \neq \emptyset$, and let W be a wellfounded relation on Z. Moreover, suppose that G is a function with values in X such that $dom(G)$ consists of all pairs of the form $\langle f, z\rangle$, where $z \in Z$ and f is a function that maps $I_W(z)$ into X.*

Then there exists exactly one function $F : Z \to X$ such that

(R) $$F(z) = G(F|I_W(z), z) \quad \textit{for each } z \in Z.$$

Theorem 26 (Principle of Generalized Recursive Constructions): *Let $X, Z \neq \emptyset$, and let W be a wellfounded relation on Z. Moreover, suppose that G^* is a function with values in $\mathcal{P}(X)\backslash\{\emptyset\}$ such that $dom(G^*)$ consists of all pairs of the form $\langle f, z\rangle$, where $z \in Z$ and f is a function that maps $I_W(z)$ into X. Then there exists a function $F : Z \to X$ such that*

(R*) $$F(z) \in G^*(F|I_W(z), z) \quad \textit{for each } z \in Z.$$

Proof of Theorem 25: Let X, Z, W, G be as in the assumptions. We first prove uniqueness of the function F. Suppose F_1, F_2 are functions such that (R) holds, and denote $C = \{z \in Z : F_1(z) \neq F_2(z)\}$. If $C \neq \emptyset$, then there exists a W-minimal element z_0 of C. For such z_0, the equality $F_1|I(z_0) = F_2|I(z_0)$ holds. Hence, $F_1(z_0) = G(F_1|I(z_0), z_0) = G(F_2|I(z_0), z_0) = F_2(z_0)$, which shows that $z_0 \notin C$. Therefore, $C = \emptyset$, which means that $F_1 = F_2$.

Now let Y be the set of all $z \in Z$ for which there exists a function $F_z : I(z) \to X$ such that

$(\mathrm{R})_z$ $$F(y) = G(F|I(y), y) \quad \text{for each } y \in I(z).$$

Note that if $z \in Y$, then $X, I(z), W \cap (I(z) \times I(z))$ together with the relevant restriction of G satisfy the assumptions of Theorem 25. Therefore, by what we have already proved, for each $z \in Y$ there exists a unique function F_z that satisfies $(\mathrm{R})_z$.

Lemma 27: *Let $z \in Z$ be such that $I(z) \subseteq Y$. Then $z \in Y$.*

Proof: For z as in the assumption, let $J = I(z) \setminus \bigcup_{y \in I(z)} I(y)$. Define $F_z = \bigcup_{y \in I(z)} F_y \cup \{\langle y, G(F_y, y)\rangle : y \in J\}$.

EXERCISE 40(PG): Show that F_z is a function with domain $I(z)$ that satisfies $(\mathrm{R})_z$.

Corollary 28: $Y = Z$.

Proof: Lemma 27 asserts that the implication ($\mathrm{Ind}_W(z)$) holds for all $z \in Z$. By Theorem 21, $Y = Z$.

Now for each $z \in Z$ define $F(z) = G(F_z, z)$.

EXERCISE 41(G): Show that the function F satisfies condition (R) of Theorem 25.

This concludes the proof of Theorem 25.[7] □

We postpone the proof of Theorem 26 until Chapter 7, and conclude the present chapter with three important applications of Theorem 25.

Definition 29: Let $\langle X, \preceq \rangle$ be a p.o. A subset $I \subseteq X$ is called an *initial segment of* $\langle X, \preceq \rangle$ (or simply an *initial segment of* X when the partial order relation is implied by context) if $\forall y \in I \forall x \in X\, (x \preceq y \rightarrow x \in I)$. An initial segment I of X is *proper* if $I \neq X$. If Y is any subset of X, then we denote the set $\{x \in X : \exists y \in Y\, (x \preceq y)\}$ by $DC(Y)$. The set $DC(Y)$ is called *the downward closure of* Y or *the initial segment of* $\langle X, \preceq \rangle$ *generated by* Y.

EXERCISE 42(G): Let $\langle X, \preceq \rangle$ be a p.o.
(a) Show that X is an initial segment of $\langle X, \preceq \rangle$.
(b) Show that for every $Y \subseteq X$, the downward closure $DC(Y)$ of Y is an initial segment of X.
(c) Show that if Y is an initial segment of X iff $DC(Y) = Y$.
(d) Show that $\langle X, \preceq \rangle$ is a w.o. and Y is an initial segment of X, then *either* $I(Y) = X$, *or* $DC(Y) = I_{\preceq}(x_0)$, where x_0 is the minimum element of $X \backslash DC(Y)$.

If $\langle X, \preceq \rangle$ is a w.o., and $Y \subseteq X$ is such that $DC(Y) \neq X$, then the minimum element of $X \backslash DC(Y)$ is denoted by $\sup^+ Y$.

Theorem 30: *Let* $\langle X, \preceq_X \rangle$ *and* $\langle Z, \preceq_Z \rangle$ *be w.o.'s. Then exactly one of the following holds:*
(a) $\langle X, \preceq_X \rangle \cong \langle Z, \preceq_Z \rangle$; *or*
(b) *There exists an* $x^* \in X$ *such that* $\langle I(x^*), \preceq_X \rangle \cong \langle Z, \preceq_Z \rangle$; *or*
(c) *There exists a* $z^* \in Z$ *such that* $\langle X, \preceq_X \rangle \cong \langle I(z^*), \preceq_Z \rangle$.

Proof: Let $\langle X, \preceq_X \rangle$ and $\langle Z, \preceq_Z \rangle$ be as in the assumptions. If one of the sets X, Z is empty, the conclusion of the theorem is trivially satisfied.

---

[7]It is perhaps not entirely clear at this stage why the definition of F given right above Exercise 41 is less "suspicious" than condition (R) of Theorem 25. Note that condition (R) contains a reference to the function F itself, whereas the latter definition of F refers to another family of functions $\{F_z : z \in Z\}$. As Russell's Paradox shows, self-reference can create problems and should be avoided whenever possible. The above proof shows how F can be defined without self-reference. If this explanation does not satisfy you, then you will have to wait until Chapter 7, where we shall point out a more tangible difference between the two descriptions of F.

So assume both X and Z are nonempty, and add a new object ∞ as a largest element to X. Define a functin G as follows: If $z \in Z$ and $f : I(z) \to X \cup \{\infty\}$, then $\langle f, z\rangle \in dom(G)$ and

$$G(f,z) = \begin{cases} \infty & \text{if } DC(rng(f)\backslash\{\infty\}) = X; \\ \sup^+(rng(f)\backslash\{\infty\}) & \text{otherwise.} \end{cases}$$

Let $F : Z \to X \cup \{\infty\}$ be the function which satisfies condition (R) of Theorem 25 for this G.

Claim 31: *$rng(F) \cap X$ is an initial segment of $\langle X, \preceq_X\rangle$.*

Proof: Suppose not, and pick some $y \in DC(rng(F) \cap X)\backslash rng(F)$. Let x be the minimal element of $rng(F) \cap X\backslash I(y)$, and let $z \in Z$ be such that $F(z) = x$. Then $z \neq z_0$.

EXERCISE 43(G): Why?

Therefore, $x = F(z) = G(F|I(z), z) = \sup^+(rng(F|I(z)) \cap X)$. In other words, $I(x) = DC(rng(F|I(z)) \cap X)$. Since $y \in I(x)\backslash rng(F)$, there exists some $x' \in rng(F|I(z))$ such that $y \prec_X x' \prec_X x$. This contradicts the choice of x as the minimal element of $rng(F) \cap X$ above y. We have proved Claim 31.

Now let us consider two cases:

Case 1: $\infty \notin rng(F)$. In this case, by Claim 31 and Exercise 42(d), the range of F is either X or $I(x^*)$ for some $x^* \in X$. Thus, the following claim shows that either (a) or (b) of Theorem 30 holds.

Claim 32: *F is order-preserving.*

Proof: Since both $\langle X, \preceq_X\rangle$ and $\langle Z, \preceq_Z\rangle$ are l.o.'s, it suffices to show that $F(z) \prec_X F(z')$ whenever $z \prec_Z z'$. So let $z \prec_Z z'$. Since $z' \neq z_0$ and $\infty \notin rng(F)$, we have $F(z') = G(F|I(z'), z') = \sup^+ rng(F|I(z'))$. Since $z \in I(z')$, also $F(z) \prec_X \sup^+ rng(F|I(z'))$. This proves Claim 32.

Case 2: $\infty \in rng(F)$. Let z^* be the $\preceq_Z$-minimal element of the set $F^{-1}\{\infty\}$.

EXERCISE 44(PG): (a) Show that $F|I(z^*)$ is an order isomorphism between $\langle I(z^*), \preceq_Z\rangle$ and $\langle X, \preceq_X\rangle$.
(b) Convince yourself that this concludes the proof of Theorem 30. □

Theorem 30 allows us to compare order types of wellordered sets.

Definition 33: Let $\langle X, \preceq_X\rangle$ and $\langle Y, \preceq_Y\rangle$ be w.o.'s, let $\alpha = ot(\langle X, \preceq_X\rangle)$ and $\beta = ot(\langle Y, \preceq_Y\rangle)$. We say that α *is less than or equal to* β and write $\alpha \leq \beta$ if there exists an order isomorphism of $\langle X, \preceq_X\rangle$ onto an initial segment of $\langle Y, \preceq_Y\rangle$.

EXERCISE 45(PG): Let α, β be as above. Show that if $\alpha \leq \beta$ and $\beta \leq \alpha$, then $\alpha = \beta$.

Definition 34: Let Z be an arbitrary set, let W be a wellfounded relation on Z, and let $\langle X, \prec\rangle$ be a strict w.o. A function $rk : Z \to X$ is called a *rank function for W with respect to* $\prec$ if $rk(z) = \sup^{+} rng(rk|I(z))$ for all $z \in Z$.[8]

EXERCISE 46(G): Let $Z = \mathcal{P}(\mathcal{P}(\mathcal{P}(\emptyset)))$. Find a rank function $rk : Z \to \omega$ for $\in$ with respect to the natural order on ω.

EXERCISE 47(G): Let $X, Z, W, \prec$ be as in Definition 34.

(a) Show that if $rk : Z \to X$ is a rank function for W, then the following holds:

$$(*_{rk}) \qquad \forall y, z \in Z\,(y \neq z \wedge \langle y, z\rangle \in W \to rk(y) \prec rk(z))$$

(b) Show that if $rk : Z \to X$ is a rank function for W and $f : Z \to X$ is any function such that

$$(*_f) \qquad \forall y, z \in Z\,(y \neq z \wedge \langle y, z\rangle \in W \to f(y) \prec f(z)),$$

then $rk(z) \preceq f(z)$ for all $z \in Z$.

(c) Show that if $rk : Z \to X$ is a rank function for W, then $rng(rk)$ is an initial segment of X.

(d) Convince yourself that if W is a strict wellorder relation on Z, then $rk : Z \to X$ is a rank function with respect to $\prec$ iff rk is an order isomorphism between $\langle Z, W\rangle$ and an initial segment of $\langle X, \prec\rangle$.

Do all wellfounded relations have rank functions? Yes. We will show in Chapter 9 that every set can be wellordered. Since $|\mathcal{P}(Z)| > |Z|$ for every Z, the existence of rank functions follows from the next theorem.

Theorem 35: *Let W be a wellfounded relation on a set Z, and let $\langle X, \preceq\rangle$ be a w.o. such that $|Z| < |X|$. Then there exists a rank function $rk : Z \to X$.*

EXERCISE 48(PG): Convince yourself that in Theorem 35, the assumption that $|Z| < |X|$ cannot be weakened to $|Z| \leq |X|$. *Hint:* Use the result of Exercise 47(d) and Example (4) from the beginning of Chapter 2.

Proof of Theorem 35: Let Z, X, W and $\prec$ be as in the assumptions. The very definition of a rank function suggests that we should use Theorem 25. Even suggestions for the choice of G are contained in Definition 34: Of course, $G(f, z)$ should be equal to $\sup^{+} rng(f)$ whenever $\langle f, z\rangle \in dom(G)$.

---

[8]If the relation $\prec$ is implied by context, then we shall just say "a rank function for W" instead of "a rank function with respect to $\prec$." As Exercise 47(b) shows, for any given strict w.o. $\langle X, \preceq\rangle$, there exists at most one rank function for W with respect to $\prec$. Once ordinals will have been defined in the framework of axiomatic set theory, we shall use the term "rank function" only for functions that take ordinals as values and are rank functions with respect to the canonical strict wellorder relation on ordinals. Therefore, from Chapter 10 on, we shall speak about *the* rank function for a wellfounded relation W. This is a matter of convenience; one doesn't have to write all the time "let $\langle X, \prec\rangle$ be a w.o." and "with respect to $\prec$." However, the ordinal concept of Chapter 10 is not a prerequisite for a rigorous treatment of rank functions.

EXERCISE 49(G): What should the domain of G be? Give a complete, formal definition of G.

Now let rk be the function F which satisfies condition (R) of Theorem 25. It follows immediately from the definition of G that rk is a rank function. So, it seems that Theorem 35 is proved. Or is it? So far, we have not yet used the assumption that $|Z| < |X|$. Since we know that this assumption cannot be dropped, the proof must be incomplete. We are in a situation that is all too familiar to every mathematician. We have an argument that looks entirely reasonable, and yet we know that it must contain a fallacy. Searching for a hidden gap in an argument can be an experience similar to looking for one's glasses: although one *knows* that they *must* be somewhere on, behind or under the desk, half an hour later they suddenly sit on the kitchen table. It is useful to know some strategies for detecting gaps in proofs. One strategy is to look for little cues like "clearly" or "obviously" that may give away hiding places of gaps.

EXERCISE 50(G): Find the cue that gives away the gap in our proof.

While this strategy helps us in identifying suspicious steps of an argument, it does not all by itself tell us whether a given step in the proof is actually wrong or merely sketchy. To find that out, one has to work with counterexamples.

EXERCISE 51(G): Consider the wellfounded relation $\leq_3$ of Example (4) from the beginning of Chapter 2, and let $Z = \mathbb{N}$, $W = \leq_3$, $\langle X \preceq \rangle = \langle \mathbb{N}, \leq \rangle$. Now try to define a function $rk_1 : Z \to X$ in exactly the same way as we defined the function rk. At what point of the recursive definition of rk_1 do things go wrong, and why?

The analysis of counterexamples shows you possible ways in which things can go wrong. Here, things can go wrong if $G(rk|I(z), z) \neq \sup^+ rng(rk|I(z))$ for some $z \in Z$. By our definition of G, this will be the case only if $DC(rng(rk|I(z))) = X$ for some $z \in Z$. When we wrote: "It follows immediately from the definition of G ... , " we forgot to check that the latter equality never holds.

Lemma 36: *$DC(rng(rk|I(z))) \neq X$ for all $z \in Z$.*

EXERCISE 52(G): Convince yourself that Lemma 36 is the last ingredient needed for the proof of Theorem 35. In other words, convince yourself that nothing else could go wrong.

Proof of Lemma 36: Suppose towards a contradiction that $C = \{z \in Z : DC(rng(rk|I(z))) = X\} \neq \emptyset$, and let z_0 be a W-minimal element of C. The function $rk|I(z_0)$ is a rank function for $W \cap (I(z_0) \times I(z_0))$.

EXERCISE 53(PG): Why?

By Exercise 47(c), $rng(rk|I(z_0))$ is an initial segment of X, i.e.,

$$X = DC(rng(rk|I(z_0))) = rng(rk|I(z_0)).$$

The latter is impossible, since $|rng(rk|I(z_0))| \leq |I(z_0)| \leq |Z| < |X|$. This contradiction proves Lemma 36, and simultaneously Theorem 35. □

Let us now look at the most important examples of wellfounded relations.

Postulate 37: *Let X be any family of sets. Then $\in$ is a strictly wellfounded relation on X.*

We call 37 a "Postulate" rather than a "Theorem," because it is difficult to imagine how it could be derived from even more elementary properties of sets. Postulate 37 is one of the axioms of ZFC, i.e., one of the sentences whose truth is taken for granted. Although we cannot prove Postulate 37, we can convince ourselves that it makes a whole lot of sense from the architect's point of view. Recall Exercise 8 and Theorem 2 of Chapter 2. As we mentioned in the Introduction, an architect of the set-theoretic universe would want to have sets assembled only from building blocks that at the time of assembly have already obtained their final shapes. Therefore, a set could not be a member of itself, nor would it be possible to have a finite sequence $\langle x_0, x_1, \ldots, x_n \rangle$ of sets so that $x_0 \in x_1 \in \ldots \in x_n \in x_0$.

EXERCISE 54(G): Convince yourself that the existence of such finite, circular sequences of sets is inconsistent with the architect's view of sets.

The existence of infinite sequences $\langle x_n : n \in \omega \rangle$ for which $x_{n+1} \in x_n$ and $x_n \neq x_m$ for all $n < m$ may look less forbidding to a truly imaginative architect. But even the most creative designer must take into account the demands of the safety inspector. Since most sets are built from infinitely many building blocks, no safety inspector could ever probe all the building blocks of the set-theoretic universe for soundness of their foundations. But the inspector could make spot checks as follows: He could randomly choose a set x_0, then randomly pick an element x_1 of x_0, then some $x_2 \in x_1$, and so on. If after finitely many steps of this process the inspector hits upon an x_n that does not require any previously assembled building blocks,[9] then he will be satisfied that the foundations of the building are sound. If not, i.e., if the inspector hits upon a sequence $\langle x_n : n \in \omega \rangle$ such that $x_{n+1} \in x_n$ for all $n \in \omega$, then our architect will be in trouble. It follows that while Postulate 37 may not be an absolute necessity for the development of set theory, it is a natural demand that should be met as a matter of prudence.

Recall Definition 4. Let Y be a transitive set, and assume that all elements of Y are sets. Then the relation $\in$ on Y is *extensional* in the following sense.

Definition 38: Let W be a binary relation on a set Y. The relation W is *extensional* if $\forall x, y \in Y \, (x \neq y \rightarrow \exists z \in Y \, ((\langle z, x \rangle \in W \wedge \langle z, y \rangle \notin W) \vee (\langle z, x \rangle \notin W \wedge \langle z, y \rangle \in W))$.

EXERCISE 55(G): (a) Show that if Y is not transitive or if Y is allowed to contain objects other than sets, then $\in$ may not be an extensional relation on Y.

(b) Show that if W is an extensional, wellfounded relation on a set Y, then Y contains exactly one W-minimal element.

[9] I.e., an x_n that is either the empty set, or an object other than a set.

(c) Show that if Y is a transitive family of sets, then $\emptyset$ is the only $\in$-minimal element of Y.

(d) Show that every strict linear order relation on any set is extensional.

Theorem 39: *Suppose W is an extensional, strictly wellfounded relation on a set Z. Then there exists exactly one transitive family of sets Y, called the* Mostowski collapse of $\langle Z, W\rangle$, *such that there exists a bijection $F : Z \to Y$ which satisfies*

(M) $$\forall z, z' \in Z\,(\langle z, z'\rangle \in W \leftrightarrow F(z) \in F(z')).$$

EXERCISE 56(G): Let a, b be any objects, and let

$$Z = \{a, \{a\}, \{\{a\}\}, \{\{\{a\}\}, a, b\}\}.$$

Find the Mostowski collapse of $\langle Z, \in\rangle$.

Proof: Let Z, W be as in the assumptions.

EXERCISE 57(PG): Show that there exists at most one transitive family Y and one bijection $F : Z \to Y$ such that (M) holds.[10]

Now let us prove that there exists at least one set Y as in the theorem. We want to use Theorem 25. Although Theorem 25 asserts uniqueness of a certain function F, this does not automatically make Exercise 57 redundant: A priori, we could have more than one choice for the function G in Theorem 25, and uniqueness of F is only guaranteed for any given choice of G. As we shall see in a moment, in the case of Theorem 39, the choice of G is practically dictated by the problem. Nevertheless, one should still carry out an independent verification of uniqueness of Y and F, since it might be possible to obtain such objects by other means than an application of Theorem 25. In order to apply Theorem 25, we need to choose X, Z, W, and G as in the assumptions of this theorem. Z and W are given, but what should X be? This is a delicate point. X must be a family of sets large enough to contain each set that may pop up in the construction. The family of all sets would do for X; but unfortunately, this is too large to form a set. The following trick allows us to overcome the obstacle. Let $\lambda = \max\{|Z|, \aleph_0\}$, and define: $H_{\lambda^+} = \{x : |TC(x)| \leq \lambda\}$. H_{λ^+} is called *the family of sets that are hereditarily of size at most* λ. In Chapter 12, we shall prove that the collection H_{λ^+} is a set.[11] Therefore, we may choose $X = H_{\lambda^+}$.

What should G be? Let $f : I_W(z) \to X$ for some $z \in Z$. If f is equal to $F|I(z)$, then $F(z) = G(f, z)$ must be the set which contains all objects of the form $f(z')$ for $z' \in I(z)$, and no other elements. In particular, if z_0 is the W-minimum element of Z (it follows from Exercise 55(b) that there exists a W-minimum element), then $G(\emptyset, z_0)$ must be $\emptyset$.

EXERCISE 58(G): Why couldn't $G(f, z)$ contain some "inessential" elements as well, for example elements that would not be in the range of the function F that we are going to construct?

---

[10] This function F will be called the *collapsing function.*

[11] By then, it will also have become clear why we adorned the symbol H_{λ^+} with a +-sign in such a strange place.

Thus, we have no other choice but to put $G(f, z) = rng(f)$. But *can* we always define G in this way? In other words, if $f : I(z) \to X$ for some $z \in Z$, is it always true that $rng(f) \in X$?

EXERCISE 59(G): Show that if λ is an infinite cardinal, $|D| \leq \lambda$, and $f : D \to H_{\lambda^+}$, then $rng(f) \in H_{\lambda^+}$.

Now let F be the function which satisfies condition (R) of Theorem 25, and let $Y = rng(F)$. It remains to verify three claims.

Claim 40: *Y is a transitive family of sets.*

EXERCISE 60(G): Prove Claim 40.

Claim 41: *F is one-to-one.*

EXERCISE 61(PG): Prove Claim 41. *Hint:* Suppose F is not one-to-one and consider the $\in$-minimal $y \in Y$ such that $F^{-1}\{y\}$ contains more than one element.

Claim 42: $\forall z, z' \in Z\, (\langle z, z' \rangle \in W \leftrightarrow F(z) \in F(z'))$.

Proof: If $\langle z, z' \rangle \in W$, then $z \in I(z')$, and $F(z) \in rng(F|I(z')) = F(z')$. If $F(z) \in F(z')$, then $F(z') = rng(F|I(z'))$, and hence $F(z) = F(z'')$ for some $z'' \in I(z')$. Since F is one-to-one, $z = z''$. It follows that $z \in I(z')$, which means that $\langle z, z' \rangle \in W$.

This concludes the proof of Theorem 39. □

EXERCISE 62(PG): (a) Find the Mostowski collapse of $\langle \omega, \leq \rangle$.

(b) Consider the relation W on ω defined by: $\langle n, m \rangle \in W$ iff $m = n + 1$. Show that W is wellfounded and extensional.

(c) Find the Mostowski collapse of $\langle \omega, W \rangle$, where W is the relation defined in (b).

Mathographical Remarks

Some authors refer to arguments like the first proof of Fact 19 as "proofs by complete induction." Moreover, induction over a structure $\langle Z, \in \rangle$ is often called "$\in$-induction." We do not need this terminology, since complete induction and $\in$-induction are just special cases of induction over a wellfounded set.

It is possible to develop a consistent axiom system for set theory in which Postulate 37 is not met. The book *The Liar* by Jon Barwise and John Etchemendy, Oxford University Press, 1987, gives a very readable introduction to this kind of set theory.

Theorem 39 plays an important role in the study of models of set theory. But in this book, we shall also use the theorem for a much more mundane purpose: In Chapter 10, an *ordinal* will be defined as the Mostowski collapse of an arbitrary w.o. This approach may raise concerns about circularity, since our proof of Theorem 39 uses the fact that H_λ is a set. This fact will be proved in Chapter 12

(Theorem 12.10), but the proof uses the ordinal concept developed in Chapter 10. While our presentation may be slightly circular, this does not reflect a real circularity in the foundations of set theory. First of all, while Theorem 39 provides motivation for the ordinal concept of Chapter 10, we shall also give an alternative definition that does not rely on Theorem 39, and yet leads to exactly the same concept. Second, it is possible to give a proof of Theorem 39 that does not require picking a set $X \supseteq Y$ prior to the construction of Y. But this technique requires some preliminaries from mathematical logic that we are going to present in the next two chapters. The alternative proof itself will be given in Chapter 7.

Part 2

An Axiomatic Foundation of Set Theory

CHAPTER 5

Formal Languages and Models

This chapter and the next are devoted to a discussion of some prerequisites from mathematical logic. They do not pretend to contain a detailed development of mathematical logic. We only hope to equip the reader with the bare necessities to make it through the rest of this book. In the present chapter, we introduce the concepts of formal languages and formal proofs, as well as of models for first-order theories. In the next chapter we show why axiomatic set theory provides a convenient foundation of mathematics, and we discuss what Gödel's Incompleteness Theorems do and do not imply about the inherent limitations of the axiomatic method.

The basic symbols of a formal language L are:

- logical connectives $\wedge$ and $\neg$;
- the quantifier symbol $\exists$;
- for each natural number i a variable symbol v_i;
- the equality symbol $=$;
- a set $\{r_i : i \in I\}$ of relational symbols;
- a set $\{f_j : j \in J\}$ of functional symbols;
- a set $\{c_k : k \in K\}$ of constant symbols;
- brackets (,),[,],{, and }.

Moreover, we have functions $\tau_0 : I \to \omega \setminus \{0\}$ and $\tau_1 : J \to \omega \setminus \{0\}$ that assign to each relational and functional symbol its *arity*.

The relational, functional, and constant symbols are called *nonlogical* symbols, the remaining ones are the *logical* symbols of the language. Depending on the particular language L, the sets I, J and K may be empty, countable or uncountable. If $I \cup J \cup K$ is a countable set, then we say that *the language L is countable.*

If you have some lingering doubts about the meaning of the symbols just introduced, rejoice. There is wisdom in this ignorance. Sit back for a moment, relax, and savor the feeling that what you have in front of you are meaningless squiggles. It will be of great help later on if you can recall your present state of mind.

Mathematical logic has two parts: the study of syntax and the study of semantics. When we study the syntax, we treat mathematical expressions just as strings of meaningless symbols. Natural languages also have a syntax, more commonly called rules of grammar. The string "every vertebrate animal moves" is a well formed expression of the English language, but the string "happy, apple, horse, happy"[1] is not. Moreover, a string like "every marsiflorous octolepian contrapidates" is also well formed according to the rules of English syntax. In Chapter 7 we shall frequently consider nonstandard models for parts of set theory. This will be rather like telling you that "every marsiflorous octolepian contrapidates" *really*

[1]This delightful example is due to Douglas Hofstadter.

means that every vertebrate animal moves, but it *could* also mean that every serious student asks questions. Think about it: It could have either of the meanings, couldn't it? Of course, the question on *your* mind now is: Wait a minute, what do you mean by *really means?* A very sensible question indeed.

Now suppose we tell you that the string "every vertebrate animal moves" also could mean that every serious student asks questions. Before dismissing the latter as utter nonsense, let us see how it differs from the previous example. Try to put yourselves in the shoes of the stranger who has mastered the rules of English grammar, who knows the meaning of "every," and also knows that "vertebrate" is an adjective, "animal" is a noun, and "moves" symbolizes some activity, but doesn't know the translation of any of these words into her native tongue. For her, the present example does not differ from the previous one. To you, the present example may look different, since you have some preconceived notion about the meaning of strings like "vertebrate." These preconceived notions sometimes get in your way if you want to study purely syntactical properties of such strings of symbols. Many students encounter this problem when studying nonstandard interpretations of familiar mathematical symbols. Whenever this happens to you, recall the above examples and see whether this helps.

EXERCISE 1(G): (a) Suppose $f^3 = x$. Find $\frac{dx}{df}$.
(b) Did you hesitate a moment while solving (a)? If so, why?[2]

It is difficult for humans to learn the grammar of a language without also learning about the meaning of at least some of its words. Therefore, we disclose the intended meaning of our symbols before describing the syntax of the language.

EXERCISE 2(G): Why did we write "intended meaning" in the previous sentence rather than just "meaning?"

As you already know, the intended meaning of the symbol $\wedge$ is "and," that of $\neg$ is "not;" $\exists$ stands for "there exists," $=$ denotes equality. The v_i's and c_k's symbolize objects in the universe of discourse. The difference is that the constant symbols stand for specific objects, whereas variables are used to refer to an object without specifying which of the objects is meant. Thus, constant symbols play the same role as proper names (like: "Hillary Clinton" or "The White House"), whereas the variables play the role of pronouns (like "it," "he," or "she." Of course, the correct word for "exists he or she" is "somebody"). The brackets delineate the ranges of logical connectives and quantifiers. They correspond to punctuation marks in natural languages.

Most of the time we shall be dealing with the language of set theory, which we denote by L_S. It has a single nonlogical symbol: the binary relational symbol $\in$. Thus, in the case of L_S, the sets J and K are empty, $I = \{0\}$, $\tau_0(0) = 2$, and for enhanced visual appeal we write "$\in$" instead of "r_0." Of course, $\in$ has the intended meaning "is an element of."

We shall also consider a language L_G that has no relational symbols, one binary functional symbol, and one constant symbol. In the case of L_G, the functional symbol f_0 will be denoted by $*$, and the constant symbol c_0 will be rendered as e.

[2]This exercise was brought to Just's attention by Walter Carlip during a lunch conversation.

For example, instead of $f_0(c_0, v_0)$ we shall write $e * v_0$. We call L_G the *language of group theory*.

An *expression* is any finite string composed of symbols of a given language; for example, both "$\exists \wedge \neg v_4 v_3$" and "$v_{12} \in v_{14}$" are expressions in the language of set theory.

Now we define the *terms*. These are the expressions that in our intended interpretation designate elements of the universe of discourse. In the case of L_S, all terms stand for sets. Let $Term_L$ (or $Term$ if the language L is implied by the context) be the smallest collection T that satisfies the following conditions:

(i) $v_i \in T$ for all $i \in \omega$;

(ii) $c_k \in T$ for all $k \in K$;

(iii) if $t_0, ..., t_{n-1} \in T$, $j \in J$ and $\tau_1(j) = n$, then $f_j(t_0, ..., t_{n-1}) \in T$.

For example, e, v_6, $e * v_0$, $(e * v_3) * e$ are all terms of L_G.

Next we define the *formulas*. These are the expressions that symbolize properties or judgements. Let $Form_L$ (or $Form$ if the language L is implied by the context) be the smallest collection F that satisfies the following conditions:

(i) if $s, t \in Term$, then $s = t \in F$;

(ii) if $t_0, ..., t_{n-1} \in Term$, $i \in I$ and $\tau_0(i) = n$, then $r_i(t_0, ..., t_{n-1}) \in F$;

(iii) if φ and ψ are in F, then so are $(\varphi \wedge \psi)$, $(\neg\varphi)$, and $(\exists v_i\, \varphi)$.

For example, $v_0 = v_1$ and $\exists v_0(v_0 \in v_1) \wedge (v_0 = v_3)$ are formulas of L_S, and $\exists v_1(v_0 * v_1 = e)$ is a formula of L_G.

Our languages do not include some popular logical connectives like $\rightarrow$ and $\forall$. This is only a matter of convenience; it allows us to reduce the number of clauses in the definitions of terms and formulas. Those extra connectives can be readily expressed by the "official" ones; e.g., if $\varphi \in Form$, then instead of $(\forall v_i \varphi)$ we can write $(\neg(\exists v_i(\neg\varphi)))$. This is unwieldy though, and the extra pairs of brackets don't exactly increase readability of the formula. Let us therefore agree on suppressing pairs of brackets whenever no ambiguity is likely to arise. In the sequel, the last expression would simply be written as $\neg\exists v_i \neg\varphi$. We also introduce the following abbreviations:

(i) $\forall v_i \varphi$ abbreviates $\neg\exists v_i \neg\varphi$;

(ii) $\varphi \vee \psi$ abbreviates $\neg(\neg\varphi \wedge \neg\psi)$;

(iii) $\varphi \rightarrow \psi$ abbreviates $(\neg\varphi) \vee \psi$;

(iv) $\varphi \leftrightarrow \psi$ abbreviates $(\varphi \rightarrow \psi) \wedge (\psi \rightarrow \varphi)$;

(v) $v_i \neq v_j$ abbreviates $\neg(v_i = v_j)$.

Moreover, in the language of set theory we use $v_i \notin v_j$ as an abbreviation for $\neg(v_i \in v_j)$.[3]

To eliminate ambiguity, we postulate a hierarchy of logical connectives: $\neg$, $\wedge$, $\vee$, $\rightarrow$, $\leftrightarrow$. Thus, $\varphi \vee \neg\varphi \wedge \chi$ is shorthand for $(\varphi \vee ((\neg\varphi) \wedge \chi))$, since $\neg$ precedes $\wedge$ and both of them precede $\vee$ in this hierarchy.

We shall frequently use x, y and z as symbols for variables. That is, each of the expressions $v_0 \in v_1$, $v_2 \in v_1$, $v_7 \in v_4$ would correctly transcribe the informal expression $x \in y$ into our formal language. Much of the time, we shall not explicitly write down a formula, not even in the semiformal language we just sketched. The reason is simple: The strings of symbols in $Form$ are not readily parsed by the

[3]In general, negation of a property will often be symbolized by a slash through the symbol for this property.

human eye. It is therefore often better to give a verbal description of a formula. If you are familiar with computers, think of our formal language as the machine language, and of verbal descriptions as a user–friendly interface. Whenever you click with your mouse on an icon on your screen, the computer executes a long series of instructions written in machine language. In much the same way, it should be possible to translate every piece of mathematical argument in this book into our formal language. The analogy fails in one important aspect, though. Working with your computer, you can instantaneously and reliably convince yourself that a given icon does correspond to an underlying piece of machine code: if not, the thing simply wouldn't run. In our case, it will take a lot of mental effort to see which human ideas can and which cannot be translated into the language of set theory. We define recursively a function $Subform$ that assigns to each $\varphi \in Form$ the set of all of its subformulas.

$$Subform(t_0 = t_1) = \{t_0 = t_1\};$$
$$Subform(r_i(t_0, ..., t_{n-1})) = \{r_i(t_0, ..., t_{n-1})\}, \text{ where } i \in I \text{ and } \tau_0(i) = n;$$
$$Subform(\varphi \wedge \psi) = \{\varphi \wedge \psi\} \cup Subform(\varphi) \cup Subform(\psi);$$
$$Subform(\neg\varphi) = \{\neg\varphi\} \cup Subform(\varphi);$$
$$Subform(\exists v_i \varphi) = \{\exists v_i \varphi\} \cup Subform(\varphi).$$

For example, the formula $\exists v_0(v_0 * v_1 = e) \vee v_1 = e$ has the following set of subformulas: $\{\exists v_0(v_0 * v_1 = e) \vee v_1 = e, \exists v_0(v_0 * v_1 = e), v_0 * v_1 = e, v_1 = e\}$.

The range of (an occurence of) a quantifier $\exists v_i$ in a formula φ is the unique subformula $\exists v_i \psi$ of φ. Thus, the range of $\exists v_3$ in $\exists v_0(v_0 \in v_2) \wedge \neg(\exists v_3(v_3 = v_1) \wedge (v_1 = v_3))$ is the formula $\exists v_3(v_3 = v_1)$. Note that our definition of $Form$ allows the same quantifier, say $\exists v_7$, to occur more than once. This is a minor nuisance of no serious consequences; but it forces us to occasionally use cumbersome phrases like: "The range of an occurence of ... ."

The occurrence of a variable v_i in a formula φ is *bound* if it happens within the range of a quantifier $\exists v_i$. Otherwise it is called *free.* Thus, in the formula $\exists v_1(v_1 = v_2) \vee \forall v_0(v_1 = v_0)$, the first occurrence of v_1 is bound, whereas the second occurrence of v_1 is free. Note that the occurrence of v_0 in this formula is bound. A formula without free variables is called a *sentence.* By $Sent_L$ (or $Sent$ if the language L is implied by the context) we denote the set of all sentences of a language L. For example, $\exists v_2(v_1 * v_1 = v_1)$ is not a sentence, but $\exists v_1(v_1 * e = v_1)$ and $e * e = e$ are sentences of L_G.

If we want to indicate that the variables $x_0, ..., x_{n-1}$ occur freely in φ, then we write $\varphi(x_0, ..., x_{n-1})$. If $t_0, ..., t_{n-1}$ are terms, then $\varphi(t_0, ..., t_{n-1})$ denotes the formula obtained from φ by substituting each free occurrence of x_i by t_i. Such a substitution is *admissible* if there is no variable z in some t_i such that x_i lies in the range of a quantifier $\exists z$. This provision insures that in our intended interpretation the formula $\varphi(x_0, ..., x_{n-1})$ makes the same assertion about $x_0, ..., x_{n-1}$ as $\varphi(t_0, ..., t_{n-1})$ makes about $t_0, ..., t_{n-1}$. For example, let $\varphi(v_0)$ be the formula $\exists v_3(v_0 \in v_1)$. Then $\varphi(v_2)$ denotes the formula $\exists v_3(v_2 \in v_1)$, and $\varphi(v_3)$ denotes the formula $\exists v_3(v_3 \in v_1)$. The former substitution is admissible, while the latter is not. In the sequel we will always tacitly assume that all the substitutions we are making are admissible. Since the notation $\varphi(x_0, ... x_{n-1})$ is primarily a convenient device to denote substitutions, we shall also use it if some of the variables x_i do not occur freely in φ, as well as in situations where not all free variables in φ are listed among the x_i's.

We also use the following abbreviations:

$\forall x_0, \ldots, x_{n-1}\varphi$ stands for $\forall x_0 \ldots \forall x_{n-1}\varphi$;
$\exists x_0, \ldots, x_{n-1}\varphi$ stands for $\exists x_0 \ldots \exists x_{n-1}\varphi$;
$\exists! x\varphi(x)$ stands for $\exists x[\varphi(x) \wedge \forall y(\varphi(y) \rightarrow x = y)]$;
$\exists x \in y\, \varphi$ stands for $\exists x\,(x \in y \wedge \varphi)$;
$\forall x \in y\, \varphi$ stands for $\forall x(x \in y \rightarrow \varphi)$.

Let $\Sigma \subseteq Form$, $\varphi \in Form$. Let us describe a notion of a *proof of* φ *from* Σ. This notion operates entirely on the level of syntax. It is modeled on a familiar experience. "I can follow this proof line by line, I understand every single step, I can vouch for its correctness, but I can't understand the idea." Who has never felt like this?

Our formal notion of proof will also have its lines and steps. A "line" is simply a formula of the language, a "step" is a mechanical rule that justifies the occurrence of line φ_n after lines $\varphi_0, \ldots, \varphi_{n-1}$. The steps are called "rules of inference." There are three demands we would like to make on the rules of inference:

Postulate 1. They should be *sound,* i.e., if all formulas in Σ as well as $\varphi_0, \ldots,$ φ_{n-1} are "true" (whatever this may mean), and φ_n is allowed by the rules as the next step in a proof, then φ_n must also be "true."

Postulate 2. They should be *effective* in the sense that some mechanical computing device could tell correct proofs from incorrect ones. Moreover, they should discriminate between correct and faulty steps only on the basis of the syntax of the formulas involved.

Postulate 3. They should be *complete* in the sense that if a formula φ can be derived by any kind of sound, purely syntactical considerations from a set of formulas Σ, then there exists a formal proof of φ from Σ.

There are sets of rules of inference that satisfy Postulates 1–3. Indeed, there are several such sets of rules. They are discussed in various textbooks on mathematical logic. It follows from Postulates 1 and 3 that none of these sets of rules is essentially better or worse than any other. If A and B are sets of rules satisfying Postulates 1–3, and if there is a proof of φ from Σ under the A–rules, there must also be a proof of φ from Σ under the B–rules.[4] To get some feel for what the rules should be like, note that any element of Σ should always be allowed as the next line of any proof. Moreover, the first line of a proof should always be either an element of Σ, or a formula that is "true" simply on the merits of its syntax. A popular example of a rule of inference is the so–called *modus ponens*. It allows a formula ψ as the next line in a proof that already contains lines φ and $\varphi \rightarrow \psi$. For instance, consider the following sequence of formulas:

$$f_j(c_0) = v_9, \;\; f_j(c_0) = v_9 \rightarrow r_4(v_9, c_0), \;\; r_4(v_9, c_0).$$

[4]There is a catch though: Not all logicians agree as to what constitutes a "sound, purely syntactical consideration." There are many sets of rules known as "nonclassical logics." These have important applications, but are not relevant for our purposes. Throughout this text, we use "classical logic," that is, our notion of a sound reasoning is the one shared by most mathematicians.

If the first two formulas are in Σ, then this is a correct proof of the formula $r_4(v_9, c_0)$ from Σ. This example also illustrates three other important features of formal proofs: proofs are *finite* sequences of formulas, the formula being proved is the last line of the proof, and the formulas occurring in proofs do not necessarily have to be sentences.

EXERCISE 3(X): Write a computer program that detects all uses of modus ponens in a formal proof.

If there exists a formal proof of φ from the formulas in Σ, then we say φ is *provable in* Σ, and write $\Sigma \vdash \varphi$. Note that $\Sigma \vdash \varphi$ may hold even if nobody has yet discovered a proof of φ from Σ. A set Σ of formulas is *inconsistent* if there is a formula φ such that $\Sigma \vdash \varphi$ and $\Sigma \vdash \neg\varphi$. Otherwise, Σ is *consistent.* The following two characterizations of consistency will be frequently used in this text without explicit reference.

A set Σ of formulas of L is consistent iff $\Sigma \nvdash \varphi$ for some sentence φ of L.

Let Σ be a set of formulas of L and let φ be a formula of L. Then $\Sigma + \{\varphi\}$ is consistent iff $\Sigma \nvdash \neg\varphi$.[5]

The kind of languages described above are *first-order languages.* They differ from higher order languages in that we are only allowed to quantify over variables, not over sets.[6] A set $T \subseteq Sent$ is also frequently called a *first-order theory.* In the sequel, the expressions "(formal) language" and "theory" should be understood to refer to first-order languages and first-order theories.

Now we turn to the discussion of semantics, that is, of how to attach meaning to the terms and formulas of a language. The language is supposed to enable us to describe certain mathematical structures called *models* of the language L or *L-structures.* Recall the description of a formal language L that we gave at the beginning of this chapter. Let A be a set. For each $i \in I$, let $R_i \subseteq A^{\tau_0(i)}$; for each $j \in J$, let F_j be a function from $A^{\tau_1(j)}$ into A; and for each $k \in K$, let C_k be an element of A. We call $\mathfrak{A} = \langle A, (R_i)_{i\in I}, (F_j)_{j\in J}, (C_k)_{k\in K}\rangle$ a *model* of the language L. The set A is called *the underlying set of* $\mathfrak{A}$, *the universe of* $\mathfrak{A}$ or *the universe of discourse.* We say that *the model* $\mathfrak{A}$ *is finite (countable, uncountable)* if the underlying set A of $\mathfrak{A}$ is finite (resp. countable or uncountable). R_i, F_j and C_k are the *interpretations* of r_i, f_j, and c_k respectively. A model for the language of set theory is a pair $\langle M, E\rangle$ with $E \subseteq M^2$. The relation E is the interpretation of the symbol $\in$.

Note that the language of set theory allows us to form expressions that will be interpreted as "for all x a certain property holds," but there seem to be no expressions that correspond to the phrase "for all *sets* x a certain property holds." The reason for this is simple: the universe of discourse of our intended interpretation of L_S consists entirely of sets, and therefore expressions like "for all sets..." are redundant. Similarly, the language of group theory L_G does not allow you to for-

[5]We write $\Sigma + \varphi$ as shorthand for $\Sigma \cup \{\varphi\}$.

[6]A brief discussion of certain limitations of the expressive power of first-order languages can be found in the next chapter.

mulate expressions of the form "for all x, y in a group G, $x * y = y * x$," but only expressions like "for all x, y, $\quad x * y = y * x$."

EXERCISE 4(G): What could the universe of discourse of a model of L_G be?

A *valuation* (of the variables) is a function s that assigns an element of A to each natural number. Given s, we assign to each $t \in Term$ some $t^s \in A$ as follows:

$$v_i^s = s(i); \qquad c_k^s = C_k;$$
$$f_j^s(t_0, ...t_{\tau_1(j)-1}) = F_j(t_0^s, ..., t_{\tau_1(j)-1}^s).$$

By recursion over the length of formulas[7] we define a relation

$$\mathfrak{A} \models_s \varphi$$

(read as: "**A** *satisfies* φ under the valuation s;" the relation $\models_s$ is called the *satisfaction relation*) as follows:

$\mathfrak{A} \models_s t_0 = t_1$ iff $t_0^s = t_1^s$;
$\mathfrak{A} \models_s r_i(t_0, ..., t_{\tau_0(i)-1})$ iff $(t_0^s, ..., t_{\tau_0(i)-1}^s) \in R_i$;
$\mathfrak{A} \models_s \neg\varphi$ iff it is not true that $\mathfrak{A} \models_s \varphi$;
$\mathfrak{A} \models_s \varphi \wedge \psi$ iff $\mathfrak{A} \models_s \varphi$ and $\mathfrak{A} \models_s \psi$;
$\mathfrak{A} \models_s \exists v_i \varphi$ iff there exists a valuation s^* such that $\mathfrak{A} \models_{s^*} \varphi$ and $s(k) = s^*(k)$ for all $k \neq i$.

Let $\varphi(v_{i_0}, ..., v_{i_{n-1}}) \in Form$ and $a_{i_0}, ..., a_{i_{n-1}} \in A$, and let s be a valuation. We write

$$\mathfrak{A} \models_s \varphi[a_{i_0}, ..., a_{i_{n-1}}],$$

if $\mathfrak{A} \models_{s^*} \varphi(v_{i_0}, ..., v_{i_{n-1}})$ for the valuation s^* that is obtained from s by replacing $s(j)$ with $s^*(j) = a_j$ for $j = i_0, \dots, i_{n-1}$ and leaving $s^*(j) = s(j)$ for $j \neq i_0, \dots, i_{n-1}$. If $\mathfrak{A} \models_s \varphi$ for every valuation s, then we write $\mathfrak{A} \models \varphi$.

When working with specific formulas, we shall often use more suggestive, self-explanatory terminology. For example, if $\mathfrak{M} = \langle M, E \rangle$ is a model of L_S and $a \in M$, then we shall write "$\mathfrak{M} \models \exists x(x \in a)$" instead of "$\mathfrak{M} \models \exists v_1(v_1 \in v_0)[a]$."

We use $\vec{a}$ as an abbreviation for finite sequences if it is clear from the context what the sequence looks like. For example, we may write $\varphi(\vec{v})$ instead of $\varphi(v_0, \dots, v_{n-1})$ and $\varphi[\vec{a}]$ instead of $\varphi[a_0, \dots, a_{n-1}]$.

EXERCISE 5(R): (a) Suppose that $\mathfrak{A}$ is a model of a given language L, and φ is a formula of that language all of whose free variables are among $v_0, \dots v_{n-1}$. Let $\vec{a} = \langle a_0, \dots, a_{n-1} \rangle$, and let s, s^* be two valuations. Show that $\mathfrak{A} \models_s \varphi[\vec{a}]$ iff $\mathfrak{A} \models_{s^*} \varphi[\vec{a}]$.

(b) Conclude that if φ is a sentence of L and $\mathfrak{A} \models_s \varphi$ for some valuation s, then $\mathfrak{A} \models \varphi$.

(c) Conclude that for any sentence φ of L, either $\mathfrak{A} \models \varphi$ or $\mathfrak{A} \models \neg\varphi$.

A *theory* in a language L is any set of sentences of this language. Let T be a theory. Then $\mathfrak{A}$ is called *a model of* T if $\mathfrak{A} \models \varphi$ for each $\varphi \in T$. We denote this property by $\mathfrak{A} \models T$.

[7]Although this type of recursion is usually referred to as "recursion over the length of formulas", it is really a recursion over the wellfounded relation of being a subformula.

For example, the models of L_G are all structures with a binary operation and a constant element. Groups are models of L_G, but so are semigroups with an identity and even structures where the operation is not associative.

Now consider the following set T of sentences in L_G:

$$T = \{\forall x, y, z\,((x * y) * z = x * (y * z)),\ \forall x\,(x * e = e * x = x),\ \forall x \exists y\,(y * x = e)\}.$$

Looks familiar? A structure $\mathfrak{A}$ is a model of T iff it is a *group,* and in the sequel we shall denote this particular theory T by GT. The three sentences in GT will be called the *axioms of* GT.

Let φ be a sentence in the language L_G. When would we call φ a theorem of GT? One way of giving precise meaning to this notion is to call φ a theorem of GT if φ is provable from the sentences in GT, i.e., if GT $\vdash \varphi$. Another, equally plausible approach is to call φ a theorem of GT if φ is true in every group, i.e., if $\mathfrak{A} \models \varphi$ whenever $\mathfrak{A} \models$ GT. Let us refer to the latter property of φ by writing GT $\models \varphi$ and saying that the sentence φ is *a consequence of* GT.

If you have studied at least some group theory, you are probably familiar with both approaches.

Let φ_0 be the sentence:

$$\forall x, y\,(y * x = e \rightarrow x * y = e).$$

This is a theorem of group theory. It asserts that the left inverse of each element is simultaneously its right inverse. Many textbooks on group theory have something like a proof that GT $\vdash \varphi_0$ near the beginning, and the reasoning involved comes about as close to a formal proof (in the sense described above) as any real proof we are aware of.

On the other hand, consider the sentences

$$\varphi_1 : \forall x, y\,(x * y = y * x)$$

and

$$\varphi_2 : \exists x, y, z \forall w\,(w = x \vee w = y \vee w = z) \rightarrow \forall p, q\,(p * q = q * p).$$

Every student of group theory learns quickly that φ_1 is not a theorem of group theory. This is shown by constructing a nonabelian group $\mathfrak{A}$, i.e., a model of GT such that $\mathfrak{A} \models \neg\varphi_1$. Thus, "not being a theorem" means "not being a consequence of GT" in this case. Later on in a standard group theory course, one learns that GT $\models \varphi_2$, a fact also referred to as "φ_2 is a theorem of group theory." The reasoning goes like this: φ_2 asserts that every group of at most three elements is abelian. It is easy enough to construct all groups of order three or less. Up to isomorphism, there are three such groups, and all of them are abelian.

But do the arguments sketched above also show that GT $\models \varphi_0$, that GT $\not\vdash \varphi_1$, and that GT $\vdash \varphi_2$?

EXERCISE 6(R): Try to show that GT $\vdash \varphi_2$, i.e., try to find a proof of φ_0 in the spirit of the familiar proof of φ_0. Be careful not to use the concept of a group (a model of GT) in your argument.

More generally, is it true that if φ is any sentence of L_G, then GT $\vdash \varphi$ if and only if GT $\models \varphi$? How about other theories? If φ is any sentence of any first-order language L, and if T is any theory in the language L, must it be the case that $T \vdash \varphi$ if and only if $T \models \varphi$?

As we have seen in the examples taken from group theory, both notions $T \vdash \varphi$ and $T \models \varphi$ formalize some aspects of our intuitive notion of φ being a theorem of T. It is therefore reasonable to expect that the two notions coincide; and in fact they do. Let us go back to Postulates 1–3 that we made when describing the rules of inference allowed in formal proofs. Postulate 1 said that these rules should be sound. If we interpret the vague notion of "true" as "$\Sigma \models \varphi$," then one can show by an easy inductive argument that soundness of our rules of inference means exactly that $T \vdash \varphi$ implies $T \models \varphi$. Therefore, the latter implication is sometimes called *the Soundness Theorem (for classical first-order logic).*

If you attempted Exercise 6, you certainly noticed that it is by no means obvious that $T \models \varphi$ implies $T \vdash \varphi$. This makes the following theorem all the more remarkable.

Theorem 1 (Gödel's Completeness Theorem, Version I): *Let φ be a sentence of a first-order language L, and let T be a theory in the language L. If $T \models \varphi$, then $T \vdash \varphi$.*

Theorem 2 (Gödel's Completeness Theorem, Version II): *Every consistent first-order theory has a model.*

EXERCISE 7(PG): Prove the equivalence of Version I and Version II of Gödel's Completeness Theorem.

EXERCISE 8(PG): Use Gödel's Completeness Theorem and the general description of the relation $T \vdash \varphi$ to prove the following:

Theorem 3 (Compactness Theorem): *If T is a theory in a first-order language L, then T has a model iff every finite subset S of T has a model.*

Let T be a theory in a first-order language L. Denote

$$T^{\vdash} = \{\varphi \in Sent_L : T \vdash \varphi\}.$$

A theory T is *complete* if $\varphi \in T^{\vdash}$ or $\neg\varphi \in T^{\vdash}$ for all $\varphi \in Sent_L$. Gödel's Completeness Theorem does *not* imply that every theory is complete: Consider our familiar example of GT. Neither φ_1 nor $\neg\varphi_1$ are in $\mathrm{GT}^{\vdash}$. The former means that the theory GT $+\neg\varphi_1$ is consistent, a fact witnessed by any nonabelian group. The latter means that the theory GT $+\varphi_1$ is consistent; models of this theory are called abelian groups. The existence of abelian groups shows that φ_1 is *consistent with* GT; the existence of nonabilian groups shows that $\neg\varphi_1$ is consistent with GT. Since both φ_1 and its negation are consistent with GT, we say that φ_1 *is independent of* GT.[8]

Later on in this text, we will encounter theorems like "CH is independent of set theory."[9] In a sense, this does not differ from the previous example. There is a set of sentences of the language L_S that we shall call "the axioms of set theory" and abbreviate by ZFC. This will be our first-order version of "set theory." CH

[8]Some authors use the expression "φ_1 is independent of GT" as a synonym for "$\neg\varphi$ is consistent with GT," and refer to the property labeled by us as "independent of GT" via the phrase "consistent with and independent of."

[9]The Continuum Hypothesis CH was briefly mentioned in Chapter 3.

can be written as a sentence of L_S. The above statement means that ZFC $\nvdash \neg$CH ("CH is consistent with ZFC") and ZFC $\nvdash$ CH ("$\neg$CH is consistent with ZFC"). On the semantic level, it means that there is a model for the theory ZFC + CH (consistency of CH) and there is a model of ZFC $+\neg$CH (consistency of $\neg$CH).[10]

Nevertheless, some mathematicians seem to have more problems with statements like "CH is independent of ZFC" than with assertions like "φ_1 is independent of GT." One apparent reason is that models of set theory are more difficult to construct than simple models for GT $+\varphi_1$. But this observation is only true to some extent. While the simplest model of ZFC $+\neg$CH is much more involved than the group $\{e\}$, it is perhaps easier to construct than the Monster Group.[11] There are deeper reasons why consistency with ZFC is somewhat special; we shall discuss them in the next chapter.

Mathographical Remarks

The Completeness Theorem was proved by Kurt Gödel in his Ph.D. dissertation at the University of Vienna in 1930.

Each of the following texts contains a detailed treatment of the material sketched in this chapter.

H. D. Ebbinghaus, J. Flum, and W. Thomas, *Mathematical Logic,* Springer-Verlag, New York, 1984.

R. E. Hodel, *An Introduction to Mathematical Logic,* PWS, Boston, 1995.

H. Enderton, *A Mathematical Introduction to Logic,* Academic Press, New York, 1972.

J. D. Monk, *Mathematical Logic,* Springer-Verlag, New York, 1976.

J. Schoenfield, *Mathematical Logic,* Addison-Wesley, Reading, Mass., 1967.

[10]Some readers may want to exclaim with horror: "But CH is only *relatively* consistent with ZFC." True enough, but did we ever claim that the statement "CH is consistent with ZFC" is true or provable? We only explained what it means.

[11]This is the largest of the so-called sporadic finite simple groups. It has $2^{46} \cdot 3^{20} \cdot 5^9 \cdot 7^6 \cdot 11^2 \cdot 13^3 \cdot 17 \cdot 19 \cdot 23 \cdot 29 \cdot 31 \cdot 41 \cdot 47 \cdot 59 \cdot 71$ elements and a very complicated structure.

CHAPTER 6

Power and Limitations of the Axiomatic Method

Since this chapter is a long one, we divide it into three sections.

6.1. Complete theories

In the previous chapter, we have seen that GT is not a complete theory in the language L_G. Are there any complete theories? Yes. Here is an example: Let TGT be the set of sentences of L_G that consists of the sentences in GT and the sentence φ_3: $\forall x\,(x = e)$.

Theorem 1: TGT *is a complete theory in* L_G.

Proof: It is not hard to see that the trivial group $\{e\}$ is the only model of TGT, and therefore TGT must be complete: If there is a sentence φ in L_G such that neither φ nor $\neg\varphi$ is in $\mathrm{TGT}^{\vdash}$, then the theories TGT $+\varphi$ and TGT $+\neg\varphi$ are both consistent. By Gödel's Completeness Theorem, there exist models $\mathfrak{A}_0 \models$ TGT $+\varphi$ and $\mathfrak{A}_1 \models$ TGT $+\neg\varphi$. These must be nonisomorphic, one–element groups, which is impossible. □

We used the word "isomorphic" already in Chapter 2. What do the notions of group isomorphism and order isomorphism have in common? Recall that a group is a model of the language L_G. If $L_{\leq}$ denotes the language whose only nonlogical symbol is the binary relational symbol $\leq$, then partial orders are models of $L_{\leq}$. We can generalize the notion of isomorphism to models of arbitrary first-order languages.

Definition 2: Let L be a first-order language as described at the beginning of Chapter 5. Let

$$\mathfrak{A} = \langle A, (R_i^A)_{i\in I}, (F_j^A)_{j\in J}, (C_k^A)_{k\in K}\rangle$$

and

$$\mathfrak{B} = \langle B, (R_i^B)_{i\in I}, (F_j^B)_{j\in J}, (C_k^B)_{k\in K}\rangle$$

be models of the language L. A *homomorphism* of $\mathfrak{A}$ into $\mathfrak{B}$ is a map $h : A \to B$ such that

(i) for each $i \in I$ and each $\vec{a} \in A^{\tau_0(i)}$, if $\vec{a} \in R_i^A$, then $h(\vec{a}) \in R_i^B$;
(ii) for each $j \in J$ and each $\vec{a} \in A^{\tau_1(j)}$, $h(F_i^A(\vec{a})) = F_i^B(h(\vec{a}))$;
(iii) for each $k \in K$, $h(C_k^A) = C_k^B$.

A homomorphism h is an *embedding*, if h is injective and satisfies the following condition (iv) which is stronger than (i):

(iv) for each $i \in I$ and each $\vec{a} \in A^{\tau_0(i)}$, $\vec{a} \in R_i^A$ iff $h(\vec{a}) \in R_i^B$.

An *isomorphism* is a surjective embedding. The models $\mathfrak{A}$ and $\mathfrak{B}$ are said to be *isomorphic* if there is an isomorphism between them. If $\mathfrak{A}$ and $\mathfrak{B}$ are isomorphic, then we write $\mathfrak{A} \cong \mathfrak{B}$. An isomorphism of $\mathfrak{A}$ onto itself is called an *automorphism* of $\mathfrak{A}$. If $A \subseteq B$, then $\mathfrak{A}$ is a *substructure* or *submodel* of $\mathfrak{B}$ if the identity function $i : A \to B$ (i.e., the function i such that $i(a) = a$ for each $a \in A$) is an embedding of $\mathfrak{A}$ into $\mathfrak{B}$.

It is common to write $\mathfrak{A} \subseteq \mathfrak{B}$ if $\mathfrak{A}$ is a substructure of $\mathfrak{B}$ and $h : \mathfrak{A} \to \mathfrak{B}$ if h is a homomorphism of $\mathfrak{A}$ into $\mathfrak{B}$, although from a strictly formalistic point of view this is wrong, since the sequence $\langle A, (R_i^A)_{i\in I}, (F_j^A)_{j\in J}, (C_k^A)_{k\in K}\rangle$ is usually neither a subset of the sequence $\langle B, (R_i^B)_{i\in I}, (F_j^B)_{j\in J}, (C_k^B)_{k\in K}\rangle$ nor the domain of h.

Example: (1) Consider the following models:
$\mathfrak{A} = \langle \mathbb{Z}, \cdot, \leq, 0\rangle$;
$\mathfrak{B} = \langle \mathbb{Z}, +, \leq, 0\rangle$;
$\mathfrak{C} = \langle \{-1, 0, 1\}, \cdot, \leq, 0\rangle$;
$\mathfrak{D} = \langle E, +, \leq, 0\rangle$.
Here $\mathbb{Z}, \cdot, +, \leq, 0$ have the standard meaning, and E is the set of even integers. Let $i_{\mathbb{Z}} : \mathbb{Z} \to \mathbb{Z}$, $i_{\{-1,0,1\}} : \{-1, 0, 1\} \to \mathbb{Z}$, and $i_E : E \to \mathbb{Z}$ be identity functions. Then $i_{\mathbb{Z}}$ is not a homomorphism of $\mathfrak{A}$ into $\mathfrak{B}$ since condition (ii) is violated. The function $i_{\{-1,0,1\}}$ is an embedding of $\mathfrak{C}$ into $\mathfrak{A}$, but not of $\mathfrak{C}$ into $\mathfrak{B}$. Thus, $\mathfrak{C}$ is a submodel of $\mathfrak{A}$, but not of $\mathfrak{B}$. The function i_E is an embedding of $\mathfrak{D}$ into $\mathfrak{B}$ (and thus $\mathfrak{D}$ is a submodel of $\mathfrak{B}$), but i_E is not an isomorphism of these two models, since it is not a surjection. However, the structures $\mathfrak{B}$ and $\mathfrak{D}$ are isomorphic: the function $h : \mathbb{Z} \to \mathbb{Z}$ defined by $h(n) = 2n$ is an isomorphism of $\mathfrak{B}$ onto $\mathfrak{D}$. The function $sgn : \mathbb{Z} \to \{-1, 0, 1\}$ that assigns 1 to each positive and -1 to each negative integer and that maps 0 to itself is a homomorphism of $\mathfrak{A}$ onto $\mathfrak{C}$. It is not an embedding, since it is not injective.

EXERCISE 1(G): (a) Convince yourself that the notions of order embedding, order isomorphism, and group homomorphism are special cases of the general notions defined above.

(b) Have you already encountered other special cases of Definition 2 in your studies of mathematics? If so, name them.

(c) Let $L_\emptyset$ be the first-order language with no nonlogical symbols whatsoever (i.e., $I = J = K = \emptyset$). Have you already encountered the notion of isomorphisms for models of this language? If so, what is it?

(d) Show that the model $\mathfrak{B}$ of Example (1) does not have any substructure whose universe is $\{-1, 0, 1\}$.

(e) Let $\langle X, \preceq_X\rangle$ and $\langle Y, \preceq_Y\rangle$ be two p.o.'s (i.e., models of the language $L_\leq$). Recall that on $X \times Y$ we can define several kinds of products, for example the simple product $\otimes^s$ and the lexicographic product $\otimes^l$. Show that if both X and Y have at least two elements, then the identity function $i : X \times Y$ is an injective homomorphism of $\langle X \times Y, \otimes^s\rangle$ into $\langle X \times Y, \otimes^l\rangle$, but not an embedding.

Let us reexamine the proof of Theorem 1. We wrote "It is not hard to see that the trivial group $\{e\}$ is the only model of TGT ..." Whenever you see a phrase like "It is not hard to see" in a mathematical text, you should switch into a mode of heightened attention. Surprisingly often, such phrases are followed by statements which are plain wrong. Let us look at the assertion we made. First of all, we should perhaps have written $\langle\{e\},*,e\rangle$ instead of $\{e\}$. This inaccuracy can be excused on the grounds that most authors write "the group G" when they mean "the group $\mathfrak{G} = \langle G,*,e\rangle$[1]." But what we wrote is wrong for a much more profound reason: Whenever $\{x\}$ is a one-element set, we can define a function $*_x : \{x\}^2 \rightarrow \{x\}$ by letting $*_x(x,x) = x$. Clearly,[2] $\langle\{x\},*_x,x\rangle$ is a group. It follows that there are infinitely many one-element groups; in fact, the collection of one-element groups even forms a proper class! When mathematicians say that there is exactly one one-element group, they mean the following:

(∗) If $\mathfrak{G}_0 = \langle G_0,*_0,e_0\rangle$ and $\mathfrak{G}_1 = \langle G_1,*_1,e_1\rangle$ are groups such that $|G_0| = |G_1| = 1$, then $\mathfrak{G}_0 \cong \mathfrak{G}_1$.

But this is expression is too unwieldy; some shorthand for it is needed. The following is a good compromise between exactitude and elegance: "Up to isomorphism, there exists exactly one one-element group."

EXERCISE 2(G): Rewrite the following claim in the spirit of (∗): "Up to isomorphism, there exist exactly two four-element groups."

Now let us take a look at the last sentence of the proof of Theorem 1. The reasoning involved in it is much deeper than it may at first glance appear.

Definition 3: Let $\mathfrak{A},\mathfrak{B}$ be two models of the same first-order language L. We say that $\mathfrak{A}$ and $\mathfrak{B}$ are *elementarily equivalent*, and write $\mathfrak{A} \equiv \mathfrak{B}$, if for every sentence φ of L, $\mathfrak{A} \models \varphi$ iff $\mathfrak{B} \models \varphi$.

Most mathematicians consider the following theorem so self-evident that they do not even notice when it is used in an argument.

Theorem 4: *Let $\mathfrak{A},\mathfrak{B}$ be two models of the same first-order language L. If $\mathfrak{A} \cong \mathfrak{B}$, then $\mathfrak{A} \equiv \mathfrak{B}$.*

Despite its being obvious, this theorem has a rather lengthy proof.

EXERCISE 3(R): Prove Theorem 4. *Hint:* Use induction over the wellfounded relation of being a subformula.

EXERCISE 4(G): Explain where in the proof of Theorem 1 a reference to Theorem 4 is hidden.

A theory that has exactly one model up to isomorphism is called *categorical.*

---

[1]In general, mathematicians are inclined to write "the model A" when they mean "the model $\mathfrak{A} = \langle A,\ldots\rangle$." This is perfectly acceptable, as long as the relations, functions and constants of the model are implied by the context.

[2]This word is also one of the cues for switching into high attention mode. But this time it precedes a correct statement.

EXERCISE 5(G): Use Theorem 4 and the technique of the proof of Theorem 1 to prove the following generalization:

Theorem 5: *Every categorical first-order theory is complete.*

EXERCISE 6(G): There exists exactly one three–element group. Find a theory T in the language L_G for which this group is the only model up to isomorphism.

EXERCISE 7(G): Show that if a theory T has an n–element model $\mathfrak{A}$ and an m–element model $\mathfrak{B}$, where n, m denote different natural numbers, then T is not complete.

EXERCISE 8(PG): Prove the following partial converse of Theorem 5:

Theorem 6: *If a complete first-order theory T has a finite model, then it is categorical.*

EXERCISE 9(PG): Use the Compactness Theorem 5.3 to prove the following:

Theorem 7: *Suppose T is a first-order theory with the following property: For every natural number n, there is a natural number $m > n$ such that T has an m–element model. Then T has an infinite model.*

Not every complete theory is categorical. Consider the language $L_{\leq}$ whose only nonlogical symbol is the binary relational symbol $\leq$. Let DLONE be the following theory in $L_{\leq}$:[3]

DLONE $= \{\varphi^i : i < 7\}$, where:

φ^0: $\forall x\,(x \leq x)$;
φ^1: $\forall x, y\,(x \leq y \wedge y \leq x \rightarrow y = x)$;
φ^2: $\forall x, y, z\,(x \leq y \wedge y \leq z \rightarrow x \leq z)$;
φ^3: $\forall x, y\,(x \leq y \vee y \leq x)$;
φ^4: $\forall x, y\,((x \leq y \wedge x \neq y) \rightarrow \exists z\,(x \leq z \leq y \wedge z \neq x \wedge z \neq y))$;
φ^5: $\forall x \exists y\,(x \leq y \wedge x \neq y)$;
φ^6: $\forall x \exists y\,(y \leq x \wedge x \neq y)$.

Of course, you recognized DLONE as the theory whose models are Dense Linear Orders with No Endpoints. It follows immediately that every model of DLONE is infinite,

EXERCISE 10(G): Why?

and that both $\langle \mathbb{Q}, \leq \rangle \models$ DLONE and $\langle \mathbb{R}, \leq \rangle \models$ DLONE. Therefore, the theory DLONE is not categorical.

EXERCISE 11(G): State the most obvious reason why $\langle \mathbb{Q}, \leq \rangle \not\cong \langle \mathbb{R}, \leq \rangle$.

Nevertheless, the theory DLONE is complete. To see this, we need the following.

---

[3]Well, in a rather informal version of $L_{\leq}$.

Theorem 8 (The (Downward) Löwenheim-Skolem Theorem): *Let L be a first-order language that contains only countably many nonlogical symbols, and let T be a theory in L. If T has any model at all, then T has a countable model.*

A proof of Theorem 8 will be given in Chapter 24. For now, let us just show why it is essential to assume that L has at most countably many nonlogical symbols.

EXERCISE 12(PG): Let $L_{\mathbb{R}}$ be the first-order language with $I = J = \emptyset$ and $K = \mathbb{R}$. Let T be the collection of sentences of the form "$c_r \neq c_s$," where r, s are different reals. Show that T has a model, and that every model of T must be uncountable.

Definition 9: Let $\mathfrak{m}$ be a cardinal, and let T be a first-order theory. We say that T is $\mathfrak{m}$-*categorical* [4] if there exists, up to isomorphism, exactly one model $\mathfrak{A}$ of T such that the universe A of $\mathfrak{A}$ has cardinality $\mathfrak{m}$.

In this new terminology, we can formulate Cantor's characterization of the order type η of $\langle \mathbb{Q}, \leq \rangle$ (Theorem 4.16) as follows:

Theorem 10: *The theory* DLONE *is* $\aleph_0$*-categorical.*

Corollary 11: DLONE *is a complete theory in* $L_\leq$.

Proof: Suppose not, and let φ be a sentence of $L_\leq$ such that neither φ nor $\neg\varphi$ is a consequence of DLONE. By Gödel's Completeness Theorem, there exist models (i.e., l.o.'s) such that $\langle A, \leq_A \rangle \models$ DLONE $+ \varphi$ and $\langle B, \leq_B \rangle \models$ DLONE $+ \neg\varphi$. Since the set $\{\leq\}$ of nonlogical symbols of $L_\leq$ is countable, the Löwenheim-Skolem Theorem applies. Therefore, we may without loss of generality assume that both A and B are countable.[5] As you saw in Exercise 10, both A and B must be of cardinality $\aleph_0$. Now it follows from Theorem 10 that $\langle A, \leq_A \rangle \cong \langle B, \leq_B \rangle$. On the other hand, since $\langle A, \leq_A \rangle \models \varphi$ and $\langle B, \leq_B \rangle \models \neg\varphi$, the two models are not elementarily equivalent. This contradicts Theorem 4. □

Recall that the idea of the proof of Theorem 1 could be expressed in the form of Theorem 5. If you compare the proofs of Corollary 11 and Theorem 1, you will see certain analogies.

EXERCISE 13(PG): Formulate and prove the analogue of Theorem 5 for $\mathfrak{m}$-categoricity.

---

[4] Some authors use the expression "categorical in power $\mathfrak{m}$" instead.

[5] This is a popular shortcut. You want to learn how to recognize it, and how to use it in your own writing. Here is what it stands for: *Since $\langle A, \leq_A \rangle$ is a model of DLONE $+\varphi$, by the Löwenheim-Skolem Theorem, there exists a countable model $\langle A', \leq_{A'} \rangle$ of DLONE $+\varphi$. Similarly, since $\langle B, \leq_B \rangle$ is a model of DLONE $+\neg\varphi$, by the Löwenheim-Skolem Theorem, there exists a countable model $\langle B', \leq_{B'} \rangle$ of DLONE $+\neg\varphi$. Instead of working with $\langle A', \leq_{A'} \rangle$ and $\langle B', \leq_{B'} \rangle$ in the remainder of the proof, let us return to its beginning; and, with the benefit of hindsight, let us choose countable models $\langle A, \leq_A \rangle$ and $\langle B, \leq_B \rangle$ right away.*

You may have already come across a version of Theorem 4 in the following context. Mathematicians often speak about *invariants.* An invariant is a property that is preserved by a certain class of isomorphisms. Order isomorphisms give rise to order invariants, group isomorphisms preserve group invariants, and topological invariants are properties preserved by homeomorphisms. Invariants are easy to recognize: If a property can be expressed in group-theoretical terms, then it is a group invariant; if it can be entirely expressed in terms of the topology, then it is a topological invariant; etc. For example, the statement: "$\forall x, y\,(x + y = y + x)$" is a group invariant, but the statement: "$\forall x, y, z\,(x < y \to x + z < y + z)$" is not. For this reason, instead of "group invariants," "order invariants" or "topological invariants," we often speak of *group properties, order properties* or *topological properties.* In particular, if a property can be expressed in a first-order language L, then it is an invariant of the class of isomorphisms of models of L. The latter observation is nothing else but Theorem 4.

Let us reexamine Corollary 11 in the light of this observation. As Exercise 4.27 shows, Dedekind completeness is an order invariant. This result conforms with our expectations, since Dedekind completeness is defined exclusively in terms of the order relation. On the other hand, the l.o. $\langle \mathbb{R}, \leq \rangle$ is Dedekind complete, while $\langle \mathbb{Q}, \leq \rangle$ is not. This means that Dedekind completeness is an order property that distinguishes the l.o.'s $\langle \mathbb{Q}, \leq \rangle$ and $\langle \mathbb{R}, \leq \rangle$. But both $\langle \mathbb{Q}, \leq \rangle$ and $\langle \mathbb{R}, \leq \rangle$ are models of DLONE, and DLONE is a complete theory. Therefore, the two models satisfy exactly the same sentences of $L_{\leq}$. In particular, any sentence φ of $L_{\leq}$ that expresses Dedekind completeness must be either true in both models or false in both of them. Seemingly, we have reached a contradiction. Where did we go wrong?

EXERCISE 14(G): Convince yourself that we have correctly shown the following: *Corollary 11 implies that no sentence φ of $L_{\leq}$ expresses Dedekind completeness.*

But didn't we also show that Dedekind completeness is an order invariant, i.e., an order property? Yes, we did. However, being expressible by a sentence φ of $L_{\leq}$ is not the same as being an order property. Let us try to write down a definition of Dedekind completeness in symbolic form. Here is one:

$$\text{(D)}\ \forall A, B\,((\forall x\,((x \in A \wedge x \notin B) \vee (x \in B \wedge x \notin A)) \wedge \forall x \in A \forall y \in B\,(x \leq y)) \\ \to (\exists z \in A \forall w \in A\,(w \leq z) \vee \exists z \in B \forall w \in B\,(z \leq w))).$$

EXERCISE 15(G): Convince yourself that this expression is indeed a definition of Dedekind completeness.

But (D) is not a sentence of $L_{\leq}$—not even of a semiformal version of $L_{\leq}$. First of all, "$\in$" is not a relational symbol of $L_{\leq}$. Second, variables of $L_{\leq}$ are supposed to symbolize only *elements* of the universe of discourse, whereas the variables A and B in (D) symbolize *subsets* of the universe of discourse. A definition that contains references to subsets of the universe of discourse as well as (possibly) to elements of the universe is called a *second-order definition.* Definitions which contain only references to elements of the universe of discourse are called *first-order definitions.* Only the latter definitions can be rendered by sentences of first-order languages. Given a first-order language L, a property P of models of this language is called a *first-order property* if it has a first-order definition. This means that there is a sentence φ of L such that a model $\mathfrak{A}$ of L has property P iff $\mathfrak{A} \models \varphi$. P is a *second-order property* if it has a second-order definition, but no first-order definition. All other properties are called *higher-order properties.*

EXERCISE 16(PG): (a) How would you describe a *third-order definition?* Model your concept and description of it on our discussion of second-order definitions.
(b) Give a mathematically meaningful example of a third-order definition.

EXERCISE 17(R): We introduced second-order properties in a rather informal way. If you want a more formal definition of second, third, fourth, ... order properties, develop your own formalism.

EXERCISE 18(G): We introduced second-order properties in a rather informal way. If you want a more formal definition of second, third, fourth, ... order properties, consult the Mathographical Remarks for a reference.

In this new terminology, we can express our musings about Corollary 11 as follows: *Since Corollary* 11 *implies that* $\langle \mathbb{Q}, \leq \rangle$ *and* $\langle \mathbb{R}, \leq \rangle$ *have the same first-order properties, it follows that Dedekind completeness is not a first-order property. The standard definition of Dedekind completeness is a second-order definition. Therefore, Dedekind completeness is a second-order property.*

Recognizing second-order properties is not always easy. To be sure, determining the order of any given definition of P is usually a straightforward business. But determining whether there exists an equivalent definition of lower order may be excruciatingly difficult. Let us consider two examples.

The notion of being isomorphic to a given structure is usually not a first order property.

EXERCISE 19(G): (a) Show that if being isomorphic to $\langle \mathbb{Q}, \leq \rangle$ were a first-order property of $L_{\leq}$, then DLONE could not have two nonisomorphic models.
(b) Why does (a) not contradict Cantor's characterization of $ot(\langle \mathbb{Q}, \leq \rangle)$?

However, there are exceptions.

EXERCISE 20(PG): Suppose $\mathfrak{G}$ is a finite group. Show that there exists a sentence φ of L_G such that for every model $\mathfrak{H}$ of L_G, $\mathfrak{H} \models \varphi$ iff $\mathfrak{H} \cong \mathfrak{G}$.

Group theorists are interested in groups called "sporadic finite simple groups." The original definition of these creatures is not at all simple, and we shall not give it here. We only tell you that it is not a first-order definition. If you are familiar with the concepts involved, you may want to verify this claim.[6]

However, a gigantic, concerted effort of many group theorists recently culminated in the theorem that there are only finitely many sporadic finite simple groups.

EXERCISE 21(G): Use this result and the previous exercise to show that there is a first-order sentence φ of L_G such that a group $\mathfrak{G}$ is sporadic, finite, and simple if and only if $\mathfrak{G} \models \varphi$.

---

[6] In fact, not even the definition of a simple group is a first-order definition.

EXERCISE 22(R): If you know some details of the classification of finite simple groups, give an estimate of the number of symbols in the sentence φ of the previous exercise.

Many important invariants in mathematics are second- or even higher-order properties. For example, even the most basic definition of topology, the standard definition of a topological space, is a second-order definition. A careful look at the notion of a homeomorphism reveals that this is not an instance of Definition 2, but a kind of second-order isomorphism. First- and second-order properties of topological spaces will be discussed in some detail in Chapter 24 and in the Appendix.

Mathematicians have a great deal of confidence in the fact that, as long as a property can be expressed in the terminology of groups, order, or topology, it is a group, order, or topological invariant. This confidence reflects the realization that Theorem 4 remains true even for higher-order sentences.[7]

The inability to express second- and higher-order concepts severely limits the usefulness of most first-order theories as a framework for doing actual mathematics. It would be wrong, for example, to assume that a textbook on group theory could in principle be written in L_G. Group theorists deal with such concepts as isomorphism, normal subgroup, and group representations; and none of these can be expressed in the first-order language L_G. If a language is to be useful as a framework for doing mathematics, it must allow us to talk about sets of elements of the universe, sets of sets, and so on. In other words, it must at least contain some form of set theory.

What else do we need? Well, we need to be able to talk about groups, p.o.'s, and other mathematical objects. As we have seen, these objects can be understood as models of certain formal languages. What do we need to talk about models? Sets, pairs, relations, functions. Is that all? Yes.

EXERCISE 23(G): If you don't believe us,[8] go back to the previous chapter and convince yourself that the notions of model and satisfaction ($\models$) were defined in set-theoretical terminology.

How about extra symbols like $*$ or $\leq$? We don't need them. Expressions of L_G can be rewritten as statements about the function that interprets $*$. Expressions of $L_\leq$ can be rendered in terms of the binary relation that interprets $\leq$. And so on. Therefore, all you need is set theory.

It seems we have discovered something very important; let us put it down in a suitable style.

Thesis 12 (First Foundational Thesis): *All mathematical concepts can be expressed in set-theoretical terminology.*

You are still not entirely convinced? Good. Neither are we. What is a "Thesis" anyway? Why didn't we write "Theorem" or "Conjecture" instead? We borrowed

---

[7]This does not necessarily mean that most mathematicians are conscious of the existence of Theorem 4 and its higher-order analogues.

[8]In mathematics, a skeptical attitude is a virtue.

the word from the name of that famous statement called "Church's Thesis[9]." The First Foundational Thesis is not a theorem, since nobody has proved it. Neither is it a conjecture; for at least two reasons. First, there is already enough compelling evidence for its truth. But the evidence is of an empirical nature and has not been confirmed by a deductive proof. Second, a conjecture is something mathematicians are trying to prove or disprove. But, while it is in principle possible that the First Foundational Thesis will be contradicted by some future evidence, it is impossible that a formal proof of it will ever be found. How come?

EXERCISE 24(G): Prove that all Great Men in History were Hoosiers.[10] *Hint:* Start out with a suitable definition of a Great Man in History.

The point is: If we want to convert the First Foundational Thesis into a First Foundational Theorem, we first need to choose a precise definition of *"all mathematical concepts."* From there on, it will be mathematics. But choosing the definition is not mathematics; it is empirical science. In the process of establishing the definition, we can take into account all existing mathematics (at least theoretically), but we can never be entirely sure that no future mathematical concept will ever transcend the definition and thus falsify the First Foundational Theorem.

In this respect, the First Foundational Thesis is similar to the findings of all natural sciences. The assertions of physicists, biologists and astronomers are based on often overwhelming empirical evidence; but all of them are in principle falsifiable without being ultimately verifiable. Newtonian mechanics had been the gospel of physicists for more than two centuries before it was falsified by an experiment and replaced by relativity theory.[11] Biologists claim that the genetic information of *all* living species is coded in the DNA. The empirical evidence for this law of nature is overwhelming, and yet—if somewhere in the depths of the jungle one could find a single living creature that passes its genetic information to its offspring by a different mechanism, the universality of the claim would be shattered.

As you will see in the next chapter, set theory can be formalized as an axiomatic theory ZFC in the language L_S. In L_S, we can talk about sets, sets of sets, and so on. Does this mean that the distinction between first-order properties and second-order properties becomes meaningless for the language L_S? No. Unfortunately, the difference between first-order, second-order, and higher-order properties does not go away. It only becomes more confusing. Sorry for that.

In explaining this delicate issue, let us restrict our attention to first- and second-order properties. Remember that we are talking about properties of models of certain theories. When we talk *about* a model of a language L, we are not usually talking in L, but in a *metalanguage.* Languages like L_G lack the expressive power to be useful as metalanguages, but L_S is just right for this purpose.[12] As long as we

[9]This is the assertion that the (formally defined) class of all recursive functions coincides with the class of all (intuitively understood) computable functions. The status of Church's Thesis is exactly the same as the status of the First Foundational Thesis.

[10]If you find this exercise distasteful, do it for Texans instead.

[11]By the way, being falsified is not the same as being useless. Practically all mechanical engineering (including the construction of your latest automobile) is still based on Newtonian mechanics. This reminds us of the relationship between naive and axiomatic set theory.

[12]A reader without any preconceived notions about the "real" meaning of the symbols $\in$ and $\leq$ will immediately notice that $L_{\leq}$ would do as well.

talk in L_S about a model, say, of L_G, telling first-order properties from second-order properties is easy.

But now comes an extra twist. Set theory can be practiced as the axiomatic theory ZFC. The theory ZFC is consistent.[13] Therefore, by Gödel's Completeness Theorem 5.2, it has a model $\mathfrak{M} = \langle M, E\rangle$.[14] We shall be able to talk about properties of $\mathfrak{M}$ in L_S. If we do this, L_S suddenly assumes two distinct functions. On the one hand, it is the language for which $\mathfrak{M}$ is a model. For any given sentence φ of L_S, either $\mathfrak{M} \models \varphi$ or $\mathfrak{M} \models \neg\varphi$. We shall refer to this role of L_S by saying that "L_S is the language of the model." On the other hand, L_S plays the role of the metalanguage.

Fortunately, every time we use L_S, we use it either as the language of the model, or as the metalanguage. It is therefore possible to keep track of the different levels at which L_S is used. Writing expressions of the language and of the metalanguage in different styles greatly facilitates keeping track of the distinctions. Whenever practicable, we shall render uses of L_S as the language of the model symbolically, and uses of L_S as metalanguage in plain English. By the way, the previous sentence belongs to the meta-metalanguage.

EXERCISE 25(PG): What level of language does the sentence immediately preceding this exercise belong to, and what is the level of this one?

Okay, okay. For the sake of sanity, we shall limit the levels of introspection by concentrating exclusively on the language of the model and on the metalanguage.

The definition of a property P of the model $\mathfrak{M} = \langle M, E\rangle$ may contain references to subsets of M. Let us restrict our attention to the case where we make just one reference to one subset A of M. If $\mathfrak{M}$ is a model of set theory, then it *may* happen that the set A corresponds to an a in M. By the phrase "A corresponds to a" we mean that $A = \{b \in M : \mathfrak{M} \models b \in a\}$.

EXERCISE 26(PG): Use the definition of the satisfaction relation $\models$ that was given in the previous chapter to show that for each a in M:

$$\{b \in M : \mathfrak{M} \models b \in a\} = \{b : \langle b, a\rangle \in E\}$$

Note that even if the elements of M are sets, nothing in our definition of a model implies that the relation E must coincide with the real membership relation. Therefore, the phrase "A corresponds to a" should not be mistaken for the phrase "A is equal to a."[15]

While it may happen that a subset A of the universe M of our model $\mathfrak{M}$ corresponds to an element a of M, this will be the exception rather than the rule. To see this, recall that there are $2^{|M|}$ subsets of M, whereas the universe of $\mathfrak{M}$ has only $|M|$ elements. So, it is unreasonable to expect that any second-order definition of a property of $\mathfrak{M}$ must have an equivalent first-order counterpart.

---

[13] We shall reexamine this claim later in this chapter.

[14] In fact, what we have to say remains true if $\mathfrak{M}$ is a model of only a fragment of ZFC, or even of an entirely different theory in L_S. But the phenomena we describe are most interesting to us for models of ZFC.

[15] There is an important class of models of (fragments of) ZFC called "standard, transitive models." For these models, and only for these, it is actually true that a subset A of M corresponds to an element a of M iff $A = a$.

Let us consider a concrete example. The Löwenheim-Skolem Theorem has the following consequence:

Corollary 13: *If* ZFC *has any model at all, then* ZFC *has a countable model.*

Now consider a countable model $\mathfrak{M} = \langle M, E\rangle$ of ZFC. By the results of Chapter 3, there exists some a in M such that $\mathfrak{M} \models |a| > \aleph_0$. Let A be the subset of M that corresponds to a. Since M is countable, so is A. But $\mathfrak{M}$ staunchly "believes" that a is uncountable. Is there a contradiction? Why should there be one? Why should a model of ZFC not be entitled to hold some politically incorrect beliefs? More concretely, the expression "$\mathfrak{M} \models |a| > \aleph_0$" stands for:

$$\mathfrak{M} \models \neg\exists f\,(f : a \to \omega \wedge \forall b, c \in a\,(b \neq c \to f(b) \neq f(c)))$$

This in turn can be translated into

(+) There is no f in M such that

$$\mathfrak{M} \models f : a \to \omega \wedge \forall b, c \in a\,(b \neq c \to f(b) \neq f(c))$$

On the other hand, our assertion that the set $A = \{b \in M : \mathfrak{M} \models b \in a\}$ is countable translates into:

(++) There exists a one-to-one function F from the set $A = \{b \in M : b \in a\}$ into ω.

The two assertions (+) and (++) are not the same. Let us ignore the questions of whether $\mathfrak{M}$'s version of ω is the real ω, and whether $\mathfrak{M}$'s idea of a one-to-one function bears any resemblance to the real thing. The important point is that (++) asserts the existence of a certain function in the real world, whereas (+) declares that no such function corresponds to an element of M. Do you see any contradiction between (+) and (++)? We don't.

EXERCISE 27(PG): In Chapter 4, we mentioned that the membership relation $\in$ is strictly wellfounded. This property of $\in$ will be reflected by the axiomatic version ZFC, i.e., if $\langle M, E\rangle \models$ ZFC, then $\langle M, E\rangle \models$ "$\in$ is strictly wellfounded." Does this necessarily mean that E has to be a strictly wellfounded relation on M whenever $\langle M, E\rangle \models$ ZFC? *Hint:* Use Theorem 2.2.

We have seen two examples of complete theories. Both are special cases of a more general phenomenon. Let $\mathfrak{A}$ be a model of a language L. By $Th(\mathfrak{A})$ we denote the set of all sentences φ such that $\mathfrak{A} \models \varphi$. We call $Th(\mathfrak{A})$ the *theory of the model* $\mathfrak{A}$ or the *theory of the structure* $\mathfrak{A}$. Note that $\mathrm{TGT}^{\vdash} = Th(\langle\{e\}, *, e\rangle)$ and $\mathrm{DLONE}^{\vdash} = Th(\langle\mathbb{Q}, \leq\rangle)$. This is no accident.

EXERCISE 28(PG): Show that a theory T in L is complete iff $T^{\vdash} = Th(\mathfrak{A})$ for some model $\mathfrak{A}$ of L.

Why bother to write down lists of sentences like DLONE? Would it not be better to just talk about "the sentences in $Th(\langle\mathbb{Q}, \leq\rangle)$?" Recall Postulate 2 made on the rules of inference. The list of sentences in DLONE is short, and so is the list of sentences in GT (at any rate, much shorter than the lists of sentences in $Th(\langle\mathbb{Q}, \leq\rangle)$ or $\mathrm{GT}^{\vdash}$). More importantly, since the sets GT and DLONE are finite, it is easy to design a mechanical procedure for checking whether or not a given sentence is on the list. The latter property is also shared by some infinite lists of sentences; we will refer to it by saying that a given list of sentences is *recursive*. If a list of sentences is recursive, then it can serve as a set of *axioms* for a theory. For example, the list ZFC of axioms of set theory is infinite, and yet recursive. Postulate 2 implies that there must be a purely mechanical way of verifying that a sentence φ_i appearing in the proof of φ by virtue of being an element of T is in fact an element of T. Recursive sets of axioms have this property.[16] We call a theory T in a language L *axiomatizable* if it has a set S of axioms, i.e., if there exists a set S of sentences of L such that $S^{\vdash} = T^{\vdash}$ and membership of a sentence in S can be determined in a purely mechanical way.

6.2. The Incompleteness Phenomenon

"..., and the more he looked inside the more Piglet wasn't there."
A. A. Milne, *The House at Pooh Corner*

We have seen that $Th(\langle \mathbb{Q}, \leq \rangle)$ is axiomatizable; how about theories of other familiar structures? Like, for instance, $Th(\langle \mathbb{N}, <, +, \times, 0, 1 \rangle)$? This is a much more difficult problem. Its solution has far-reaching consequences for the foundations of mathematics, including the development of axiomatic set theory.

Let L_A be the language whose only nonlogical symbols are the binary relational symbol $<$, the binary functional symbols $+$ and $\times$, and the constant symbols $0, 1$. We call a sentence φ of L_A a *truth about arithmetic* if $\langle \mathbb{N}, <, +, \times, 0, 1 \rangle \models \varphi$, and we call φ a *falsity about arithmetic* if $\langle \mathbb{N}, <, +, \times, 0, 1 \rangle \models \neg\varphi$.

Theorem 14 (Gödel's First Incompleteness Theorem, Version I): *Let T be an axiomatizable theory in L_A such that $T \subseteq Th(\langle \mathbb{N}, <, +, \times, 0, 1 \rangle)$. Then T is not complete.*

This theorem tells us that it is impossible to discover all true statements about the arithmetic of natural numbers in a purely mechanical way. In other words, if we make a list Σ of sentences in the language L_A, then at least one of three things must happen:

(++) Σ does not contain all arithmetical truths, or
(++) there is no mechanical procedure that allows us to verify which sentences are on the list Σ, or
(++) the list Σ contains a falsity about arithmetic.

Not surprisingly, Gödel's Incompleteness Theorems (we shall encounter the second of them in a moment) have ignited the imagination of mathematicians, philosophers, writers and journalists. There are some highly controversial interpretations of these theorems. We welcome the interest of the wider public and we respect intellectual freedom. However, since Gödel's theorems are germane to our development of set theory, we must apply the standards of rigor characteristic of our discipline. This means we have to very carefully distinguish the mathematical facts from philosophical speculation whenever we invoke one of Gödel's Incompleteness Theorems.

[16]It suffices to require that the set of axioms is *recursively enumerable* rather than recursive. This subtlety is irrelevant for the development of set theory and will be ignored.

To illustrate this point, let us consider in detail one of the popular interpretations of Theorem 14.

Gödel's First Incompleteness Theorem, Tabloid Version:

FAMOUS LOGICIAN PROVES:

(!!!) *We shall never know the whole truth about arithmetic.*

Is that so? This depends on how one interpretes the tabloid headline. If (!!!) is only supposed to mean that at any given time in the future some arithmetical truth will not yet have been pronounced true by any mathematician, then (!!!) is of course a correct claim. But you do not need the genius of Kurt Gödel to figure that out. This reading of (!!!) simply follows from the fact that there are infinitely many truths about arithmetic,

EXERCISE 29(G): Why are there infinitely many truths about arithmetic?

while there will have been only finitely many mathematicians living up to any given moment in the future, and each of these mathematicians will have been able to prove only finitely many sentences of L_A during a lifetime. So there will always be infinitely many yet unproven truths.

EXERCISE 30(G): Before reading on, try to find a flaw in the above interpretation.

Our argument makes sense only if we adopt a grimly authoritarian point of view of what has and what has not been proven to date. "Been proven" would have to mean something like "been explicitly asserted by some mathematician, who also supplied a conclusive argument of its truth." Isn't that what "been proven" means? Well, let ψ_0 be the sentence "Every even number greater than two is composite."

EXERCISE 31(G): (a) Rewrite ψ_0 in the language L_A without using abbreviations.
(b) Express in L_A: "4 is a composite number."

Suppose ψ_0 has been pronounced true by the venerable ancient XYZetos in his famous Tractatus Veritates Ultimatis. But has XYZetos also asserted that 4 is a composite number? Perhaps so. How about 444?

EXERCISE 32(X): Express the sentence "444 is a composite number" in L_A. Remember that you are not allowed to use such abbreviations as "4."

Done with Exercise 32? Good. So you will probably agree that even with a gappy knowledge of the ancient writings, it is rather safe to assume that the famous Tractatus does not contain the explicit assertion that 444 is a composite number. Or does it? Isn't the explicitly given proof of ψ_0 enough of an explicit assertion that 444 is composite? Perhaps so, but then we also have to give XYZetos credit for proving that 1234567890 is composite, and for infinitely many more theorems as well.

Oops, what about our assertion that every mathematician can only assert finitely many theorems during a lifetime? You can't have it both ways. Either you imagine whole academies full of ancients busily producing lists of theorems like: "4 is a composite number. 6 is a composite number. 8 is a composite number. 10 is a composite number. ... ," or you adopt a more relaxed version of "Having proved a theorem" and admit the possibility of an infinitely productive individual. The latter view is more in line with mathematical practice; so let us adopt it. The old counting argument given before Exercise 30 is incompatible with our new liberal view of what constitutes a proven theorem.

This reopens the question: Is it true that we shall never know all truths about arithmetic? To answer it, we must once more reexamine our concept of "knowing" a particular truth about arithmetic. Why are we ready to give XYZetos credit for asserting that 1234567890 is composite; although this number appears nowhere in the Tractatus? One is tempted to say that the latter is a trivial consequence of XYZetos' theorem ψ_0, but the temptation to use the word "trivial" ought to be resisted whenever humanly possible. For those who cannot resist, we recommend the following exercise.

EXERCISE 33(X): (a) Define your own favored concept of triviality. It should be somewhat reasonable, objectively verifiable, and mathematically rigorous.

(b) Find two sentences ψ_1 and ψ_2 such that ψ_2 follows trivially from ψ_1, but every proof you can think of does not want to conform to your definition worked out in point (a).

So, let us take a second look at our hypothetical ancients producing the infamous list of theorems: "4 is a composite number. 6 is a composite number. 8 is a composite number. 10 is a composite number. ..." What makes them ridiculous is not only the triviality of their pursuit, but also the purely mechanical aspect of their activity. In a sense, the great XYZetos laid down a rule for producing a list of theorems in a purely mindless way. In ancient times, the latter thankless task could be conveniently left to slaves; in modern times we have computers for that same purpose. So we might say that a truth about arithmetic is *implicitly* established, if it can be derived in a purely mechanical way from truths already explicitly asserted by mathematicians. Thus, XYZetos would have explicitly asserted that every even number is composite, and implicitly, that 444 is composite. But what does "in a purely mechanical way" mean? A computer cannot think; it can only carry out instructions according to specific rules. So "mechanical" should mean "according to specified rules." Just any kind of rules? No. A rule like "Whenever φ has been established as a truth, $\neg\varphi$ can be added to the list of mathematical truths" does not serve our purposes well.

EXERCISE 34(G): What properties should the rules for deriving further mathematical truths from already established ones have?

Are you done with Exercise 34? Does the list of properties you have come up with look familiar? Congratulations! You have reinvented the idea of a set of rules of inference. Now we can give a precise meaning to the notion of "being implicitly established": Let ARI(t) be the set of arithmetic truths that have been explicitly established by time t in history. Then we say that a sentence φ of L_A has been *implicitly established* by time t iff $\varphi \in \mathrm{ARI}(t)^{\vdash}$. Note that while ARI($t$) changes with the publication of every new result in number theory, $\mathrm{ARI}(t)^{\vdash}$ may remain the same over long periods of history. $\mathrm{ARI}(t)^{\vdash}$ changes only when we accept new axioms.

Equipped with this concept, let us return to the question: Is it true that we shall never know all truths about arithmetic?

Again, picture a moment in history, and all the mathematicians that (will) have lived up to this moment. Each of these mathematicians will have contributed a finite number of mathematical insights to the study of arithmetic. "An insight" could be either establishing the truth of one sentence of L_A, or giving a mechanical rule for writing out a list of true statements of L_A.[17] Hence, the combined wisdom of all mathematicians who will have lived until that specific point in history can be summed up in a finite number of effectively verifiable lists of true statements of L_A. The union of these lists is again an effectively verifiable list of sentences of L_A, so it may serve as a set of axioms. Let us assume the moment in history we are looking at is the year 6991, and let us denote this set of axioms by ARI(6991). Making the unrealistic assumption that none of the mathematicians living until 6991 will have bequeathed a false theorem to posterity, we see that

$$\mathrm{ARI}(6991) \subseteq Th(\langle \mathbb{N}, <, +, \times, 0, 1\rangle).$$

Since our rules of inference are sound, also

$$\mathrm{ARI}(3991)^{\vdash} \subseteq Th(\langle \mathbb{N}, <, +, \times, 0, 1\rangle).$$

Now the question of whether in 6991 we shall know all truths about arithmetic can be reformulated as:

$$\text{Is } \mathrm{ARI}(6991)^{\vdash} = Th(\langle \mathbb{N}, <, +, \times, 0, 1\rangle)?$$

To this question, Gödel's First Incompleteness Theorem does give a negative answer. And of course, there is nothing special about the year 6991. Thus, we can indeed conclude that at any given time in history there will be arithmetical truths not yet implicitly established.

Our concept of "having been implicitly established" is not a very good formalization of the intuitive concept of "having been proved." Note that if we write $\varphi \in \mathrm{ARI}(1996)^{\vdash}$, then this does not mean that φ follows trivially from what is known these days. It only means that there exists some proof from currently accepted axioms, no matter whether this proof has already been discovered by somebody, or whether it is likely to be discovered soon. For example, Fermat's Last Theorem had been implicitly established even before Fermat made the famous claim that he had proved it. But it was proved only two months before this sentence was written. The author of the proof, Andrew Wiles, did not lay down new axioms. Neither did he discover new rules of inference. There exists a proof of Fermat's Last Theorem

[17] The latter also looks like a rule of inference, but there is a subtlety which for now can be explained as follows: Our hypothetical XYZetos could also have had some interests in, say, geometry. One of XYZetos' geometrical insights might yield an algorithm for computing $\sqrt{2}$ with any desired accuracy. This algorithm would in effect lead to a list of arithmetical statements, like: "$1 \times 1 < 2$, $1.4 \times 1.4 < 2$, $1.41 \times 1.41 < 2, \ldots$" (Decimal fractions are not part of L_A, but can be easily translated into L_A.) However, the procedure involved does not have to be describable by a finite number of rules of inference. The rules of inference lead from formulas of L_A to formulas of L_A, and XYZetos' algorithm was supposed to originate from a theorem in geometry. So, apart from rules of inference in the strict sense, there might also be "axiom-generating rules." Such axiom-generating rules are called *axiom schemata*.

in ZFC that uses only rules of inference which can be looked up in any introductory textbook on mathematical logic.[18] In principle, such a proof could have been discovered by a computer that runs long enough and has enough memory. The problem is that such a proof would be horribly long and totally indigestible by any human reader. Also, it is unlikely that even the best available computer could finish this job within the next few billion years.[19] Andrew Wiles' contribution to mathematical knowledge consists of discovering some clever shortcuts that make the proof fit on a mere 200 pages. In this form, the proof can be understood and verified by at least some humans.

As the above example and Exercise 33 show, our intuitive idea of when a theorem has been proved is too elusive to be captured in a formal definition. It can only be approximated. "Being explicitly stated and proved in the mathematical literature" is an approximation from below; "being implicitly established" is a *very* rough approximation from above.

So far, so good. But the tabloid version of Gödel's theorem has a certain mysterious ring to it, doesn't it? This sense of mystery is totally absent from the interpretations we have given so far. Of course, creating a false sense of mystery is exactly what tabloids are all about; but aren't we perhaps missing something? What, precisely, accounts for the mysterious air of the tabloid version? Well, it sounds as if Gödel had proved that there is some sentence of L_A whose truth or falsity can never be established. Has he done that?

In 1889, the Italian mathematician Giuseppe Peano formulated the axioms of a theory PA in L_A.[20] The axioms of PA are evidently arithmetic truths. Although the theory PA is remarkably simple, all theorems of arithmetic that were known to Gödel's contemporaries are theorems of PA. In other words, before 1931, the year in which Gödel published his Incompleteness Theorems, the collection of "implicitly established truths about arithmetic" was synonymous with the set $PA^{\vdash}$. If we want to understand how much truth there is to the mysterious aspect of (!!!), we need to answer the following questions.

Question 1: Did Gödel prove that there is a sentence φ of L_A such that neither $PA \vdash \varphi$ nor $PA \vdash \neg\varphi$?

Question 2: Did Gödel point out a specific sentence φ of L_A such that neither $PA \vdash \varphi$ nor $PA \vdash \neg\varphi$?

Question 3: Did Gödel prove that there is a sentence φ of L_A such that the truth or falsity of φ could never be determined by any means whatsoever?

Question 4: Did Gödel point out a specific sentence φ of L_A such that the truth or falsity of φ could never be determined by any means whatsoever?

The tabloid headline could refer to any of these four accomplishments. If the answers to Questions 3 or 4 turn out to be positive, the mystery will still be there,

---

[18] Of course, we know of the existence of such a proof only thanks to Wiles' result.

[19] We are not kidding.

[20] The letters stand for "Peano Arithmetic."

despite its elevation from the level of the "National Inquirer" to that of the "American Mathematical Monthly." Let us try to answer these four questions.

EXERCISE 35(G): Show that the answer to Question 1 is positive.

We should mention here that Gödel's First Incompleteness Theorem is usually stated in a different form.

Theorem 15 (Gödel's First Incompleteness Theorem, Version II): *Let T be a consistent, axiomatizable theory in L_A such that* PA $\subseteq T$. *Then T is not complete.*

This version gives slightly more information than Version I, since the theory $Th(\langle \mathbb{N}, <, +, \times, 0, 1\rangle)$ is only one of many consistent extensions of PA. However, Version I better reflects the philosophical significance of the theorem. While PA can be considered an accidental creation of the human mind, $Th(\langle \mathbb{N}, <, +, \times, 0, 1\rangle)$ seems to exist independently of human intelligence.[21]

The answer to Question 2 is also positive, but it does not follow from Gödel's First Incompleteness Theorem. It is a consequence of a more specific statement.

Theorem 16 (Gödel's Second Incompleteness Theorem): *Let T be an axiomatizable first-order theory in a formal language L such that T has at least the expressive power of* PA. *Then there exists a sentence $CON(T)$ of L that asserts the consistency of T and such that, unless T is inconsistent, $CON(T)$ is not provable in T.*

We do not wish to define exactly what it means for a theory T to have "at least the expressive power of PA." For our purposes it suffices to know that this concept does have a precise definition, and that PA, ZFC, and all axiomatizable extensions of PA or ZFC qualify. A trickier part of Theorem 16 is the phrase "a sentence $CON(T)$ that asserts the consistency of T." If $L = L_S$, and T is ZFC, then we can imagine that $CON(\text{ZFC})$ says: "There exists a model $\langle M, E\rangle \models$ ZFC." But what about PA? The sentences in L_A do not speak about consistency. They speak about numbers, addition, and multiplication, don't they? They don't. As we pointed out in Chapter 5, expressions of a given formal language L do not speak about anything. Meaning is in the mind of the beholder. Gödel's proof shows us how to construct a sentence $CON(\text{PA})$ of L_A which, in the standard interpretation of the symbols, "speaks" about some obscure properties of prime numbers. It also shows how we can, by attaching nonstandard meanings to the symbols of $CON(\text{PA})$, convince ourselves that $CON(\text{PA})$ is an arithmetical truth if and only if PA is a consistent theory. Moreover, Gödel's proof shows that PA is inconsistent if and only if PA $\vdash CON(\text{PA})$. Similar sentences can be constructed for ZFC and other axiomatizable theories with sufficient expressive power.

This result gives a positive answer to Question 2. Does it also give a positive answer to Question 4? No. The sentence $CON(\text{PA})$ is true. How do we know? For one thing, it can be proved in ZFC that PA is a consistent theory. More

[21] Depending on your world view, you may interprete this as "$Th(\langle \mathbb{N}, <, +, \times, 0, 1\rangle)$ is God's creation," or "Extraterrestrial intelligent beings will probably be interested in $Th(\langle \mathbb{N}, <, +, \times, 0, 1\rangle)$, but they will probably use an axiom system different from PA."

importantly, $CON(\mathrm{PA})$ has all the trappings of a good new axiom for arithmetic. To understand this point, let us ponder the question: What makes a sentence a good candidate for an axiom? The answer is very simple: It should be recognized by practically all mathematicians as expressing a self-evident truth. It is almost impossible to imagine that all axioms of PA express truths about arithmetic, while PA is an inconsistent theory. Therefore, Gödel has not shown us a sentence whose truth or falsity could not be decided by any means whatsoever. He has shown us how to construct a *true* sentence that cannot be deduced from a given set of axioms.

What is the significance of Gödel's Incompleteness Theorems for the foundations of mathematics? Consider the following thesis.

Thesis 17 (Second Foundational Thesis): *All truths about sets that mathematicians can be fully certain of are theorems of* ZFC.

Let SET(1996) be the collection of all truths about sets that had been implicitly established by the time this book appeared. In this terminology, Thesis 17 could be expressed as follows:

Thesis 17′ (Second Foundational Thesis): $\mathrm{SET}(1996) = \mathrm{ZFC}^{\vdash}$.

Anecdotal evidence suggests that most mathematicians accept the Second Foundational Thesis. The widely held opinion that "ZFC is the foundation of all mathematics" is nothing else but a conjunction of the First and Second Foundational Theses.

However, the Second Foundational Thesis is wrong! Gödel's Second Incompleteness Theorem tells us that there is a sentence $CON(\mathrm{ZFC})$ of L_S which is

(a) true if and only if ZFC is consistent, and

(b) not in $\mathrm{ZFC}^{\vdash}$.

We have yet to meet a mathematician who has complete confidence in the axioms of ZFC, and simultaneously seriously doubts the truth of $CON(\mathrm{ZFC})$.

What is going on here? Are most mathematicians unaware of Gödel's Second Incompleteness Theorem? No, the theorem is widely known. Or are mathematicians unreasonable people? Not usually. The point is that the mathematical enterprise demands a fixed, universally accepted axiom system Σ as its foundation. Ideally, such an axiom system Σ would have to meet three demands. It should be:

(1) *Sound* in the sense that it contains only true sentences;

(2) *Effective* in the sense that axioms can be recognized by a purely mechanical procedure;

(3) *Complete* in the sense that all self-evident truths are in $\Sigma^{\vdash}$.

Gödel's Incompleteness Theorems tell us that it is impossible to have such an axiom system. The moment you settle for a Σ that satisfies (1) and (2), a sentence $CON(\Sigma)$ pops up which spoils (3). If you make Σ inclusive enough to catch all potential spoilers, you loose either soundness (1), or effectiveness (2). This is life.

EXERCISE 36(G): What can you do?

Since the ideal axiom system does not exist, you have to settle for something less. You cannot sacrifice soundness or effectiveness; so the only possibility is giving up (3). In other words, you have to draw the line somewhere. This will leave some obvious truths outside of $\Sigma^{\vdash}$, and you will perhaps have to pretend that you are "not entirely confident about the truth" of these outsiders. But as long as all really important mathematical theorems end up inside $\Sigma^{\vdash}$, you will have created an acceptable foundation of mathematics.

When we wrote that mathematicians *accept* (rather than believe in) the Second Foundational Thesis, we meant that mathematicians have reached a consensus to draw the line of complete confidence around the theorems of ZFC. In this way, our discipline acquired a healthy, if somewhat arbitrary, foundation.

Is one allowed to use such evident truths as the consistency of ZFC in a mathematical argument? One may do so, but each use of an assumption that transcends ZFC has to be explicitly acknowledged. Earlier in this chapter, we wrote unblushingly: "The theory ZFC is consistent." (Well, we did blush a little, i.e., we left a footnote.) Anyway, you should never write like this. You are supposed to write: "Let us assume that the theory ZFC is consistent." Moreover, if you made this assumption[22] in the proof of a theorem, you have to add it explicitly to the assumption of the theorem.

Let us illustrate this technique by an example. In Chapter 3 we mentioned that Paul Cohen had shown that "the Continuum Hypothesis could not be proved unless current foundations of mathematics are changed." More precisely, he has shown that CH is not a theorem of ZFC, i.e., that the theory ZFC + ¬CH is consistent. Could this be a theorem of ZFC? No. Cohen's theorem translates into the sentence $CON(\text{ZFC} + \neg\text{CH})$, and the latter implies $CON(\text{ZFC})$, which, by Gödel's Second Incompleteness Theorem, is not a theorem of ZFC. What Cohen did was the following: *He assumed that* ZFC *was a consistent theory.* By Gödel's Completeness Theorem, this implied the existence of model $\mathfrak{M}$ of ZFC. Cohen did not know whether or not $\mathfrak{M}$ would satisfy CH. But he invented a method, called the method of forcing, that allowed him to take any model $\mathfrak{M}$ of ZFC, and to transform this model into another model $\mathfrak{M}[\mathbb{G}]$ that would satisfy both ZFC and ¬CH. In other words, Cohen proved the following:

Theorem 18: *If* ZFC *is a consistent theory, then so is the theory* ZFC + ¬CH.

Symbolically, his theorem can be rendered as follows:

Theorem 18′: $CON(\text{ZFC}) \to CON(\text{ZFC} + \neg\text{CH})$.

Another popular way of expressing the same theorem is the following:

Theorem 18″: ¬CH *is* relatively *consistent with* ZFC.

It should be pointed out that the proof of Theorem 18 can be carried out entirely in ZFC. If you ever prove a theorem like Theorem 18, you can formulate it in any of these three styles. But remember that you are not allowed to just write: "Theorem: $CON(\text{ZFC} + \neg\text{CH})$."

[22]Or, more accurately: If you used this fact and want to pretend dutifully that you aren't sure about it.

EXERCISE 37(G): In Chapter 3, we also mentioned that Gödel had proved that CH cannot be disproved unless current foundations of mathematics are changed.

(a) Make a guess how the general outline of his proof might look.

(b) Render this theorem of Gödel in the styles of Theorem 18, Theorem 18′, and Theorem 18″.

Apart from additional assumptions that are true (like CON(ZFC)), you may also use assumptions that are merely relatively consistent with ZFC, like CH or its negation. One does this by saying at a suitable place in the proof: "Assume that CH holds," and by acknowledging the use of CH in the formulation of the theorem. For example like this:

Theorem 19: *Assume that* CH *holds. Then there exists a Luzin set.*

Or more parsimoniously:

Theorem 19′ (CH): *There exists a Luzin set.*

In Chapter 17, we shall define the notion of a Luzin set, and we shall also prove Theorem 19. For now, let us ponder the question: What is a theorem like Theorem 19 good for? Whereas CON(ZFC) is generally believed to be true; mathematicians are divided about the truth of CH. Some think it is true, some think it is false, some think we will eventually find means of deciding the truth or falsity of CH (such means must transcend ZFC, of course), some think that the truth or falsity of CH will remain forever undecided, and some think that it simply doesn't make sense to say that CH is "really" either true or false. We do not want to advocate any one of these points of view. You will have to work out your own position, and this may take a lifetime of mathematical research.

But there is a definite meaning of Theorem 19: The theorem implies that the theory "ZFC + There exists a Luzin set" is relatively consistent with ZFC. In other words, suppose you find a proof (in ZFC) of the nonexistence of a Luzin set. Theorem 19 tells you how to conclude from your proof (in ZFC) that CH fails. In other words, you will have proved the negation of CH in ZFC. Hence, you will have shown that ZFC + CH is an inconsistent theory. But Gödel's theorem $CON(\text{ZFC}) \rightarrow CON(\text{ZFC} + \text{CH})$ (you got this as the solution of Exercise 37(b), didn't you?) tells us that if the theory ZFC + CH is inconsistent, then so is the theory ZFC. Therefore, every proof of the the nonexistence of a Luzin set automatically proves the inconsistency of ZFC.

Or, let's put it this way: Trying to prove the nonexistence of a Luzin set is as hopeless an enterprise as trying to prove the inconsistency of ZFC.

We have not yet answered Question 3. Mathematicians seem to be somewhat divided on whether or not this is a legitimate interpretation of Gödel's Incompleteness Theorem. We leave you on your own with this question. Remember, the answer is not due at the final exam for this course. You may arrive at it at any time during your life. Or you may just not strive for an answer and savor the mystery instead. If you do want to get started at the question right now, we recommend the following exercise.

EXERCISE 38(PG): Write a short essay on the following interpretation of Gödel's Incompleteness Theorems:

Gödel's Incompleteness Theorems, EIMS[23] Version: *Mathematicians will never be entirely replaced by computers.*

We have not yet discussed one of the most puzzling aspects of Gödel's Theorem. Here it is: It follows from Gödel's Completeness theorem that if $CON(\text{ZFC})$ is not provable in ZFC, then the theory ZFC $+\neg CON(\text{ZFC})$ has a model $\langle M, E\rangle$. One can show that a theory T in a language L is inconsistent if and only if *every* sentence of L is a theorem of T. Therefore, if $\langle M, E\rangle \models$ ZFC $+\neg CON(\text{ZFC})$, then $\langle M, E\rangle \models$ "ZFC $\vdash 0 = 1$." (In the language of ZFC, "$0 = 1$" is of course the sentence "$\emptyset = \{\emptyset\}$.") In other words, $\langle M, E\rangle \models$ "There exists a formal proof of the sentence '$\emptyset = \{\emptyset\}$' from the axioms of ZFC." Recall that a formal proof is a finite sequence of formulas of L_S such that each formula in the sequence follows from the preceding ones by a simple, well-defined rule of inference. Such a sequence is a formal proof of $\emptyset = \{\emptyset\}$ if and only if its last formula is "$\emptyset = \{\emptyset\}$." The formulas appearing in a proof are not sets, but can be coded as sets. It follows from the First Foundational Thesis that there is a formula $\psi(p)$ of L_S with one free variable p which can be interpreted as "p is a formal proof of the sentence '$\emptyset = \{\emptyset\}$' from the axioms of ZFC." If $\langle M, E\rangle \models \exists p\, \psi(p)$, then there is some p in M such that $\langle M, E\rangle \models \psi(p)$, i.e., $\langle M, E\rangle \models$ "p is a formal proof of the sentence '$\emptyset = \{\emptyset\}$' from the axioms of ZFC." Although the formulas appearing in p will be coded as certain sets, and $\langle M, E\rangle$'s version of the membership relation will be E, it should in principle be possible to decode all these objects and to obtain from p a finite string of formulas which is a proof of "$\emptyset = \{\emptyset\}$" from the axioms of ZFC ...

This is a mathematician's nightmare.

In the next section, we shall restore your piece of mind by showing the flaw in the above argument.

6.3. Definability

In Chapter 5, we defined a language L_S whose only nonlogical symbol is $\in$, and called it "the language of set theory." However, in this book, we use a lot more nonlogical symbols than just $\in$. You have seen $\subseteq, \langle a, b\rangle, |a|, \dots$, and this is just the beginning. Now we are going to explain how the use of all these extra symbols can be reconciled with the notion that, in principle, we could work in the language L_S.

Consider a fixed language L, let $\varphi(v_0, ..., v_{n-1}) \in Form_L$, and let $\mathfrak{A}$ be a model of L. We let

$$\varphi^{\mathfrak{A}} = \{\langle a_0, \dots, a_{n-1}\rangle \in A^n : \mathfrak{A} \models \varphi[a_0, \dots, a_{n-1}]\}.$$

The formula $\varphi(v_0, ... v_{n-1})$ thus defines an n-ary relation $R_\varphi = \varphi^{\mathfrak{A}}$ on the universe A of $\mathfrak{A}$.

Similarly, let Σ be a theory in L, and let $\psi(v_0, ..., v_m) \in Form_L$ be such that $\Sigma \vdash \forall v_0 ... v_{m-1} \exists! v_m \psi(v_0, ..., v_m)$. Then $\psi^{\mathfrak{A}}$ corresponds to an m-ary function F_ψ on the universe of $\mathfrak{A}$: If $\langle v_0, \dots, v_m\rangle \in \psi^{\mathfrak{A}}$, then $F_\psi(v_0, \dots v_{m-1}) = v_m$.

[23]This stands for the title of the periodical "Employment Information in the Mathematical Sciences."

Fix φ and ψ as above, and let L' be the language that one obtains by adding to L two nonlogical symbols: a new n-ary relational symbol r_φ and a new m-ary functional symbol f_ψ. Moreover, let

$$\varphi^* \equiv \forall v_0 ... v_{n-1}(r_\varphi(v_0, ..., v_{n-1}) \leftrightarrow \varphi(v_0, ..., v_{n-1}));$$
$$\psi^* \equiv \forall v_0 ... v_m(f_\psi(v_0, ..., v_{m-1}) = v_m \leftrightarrow \psi(v_0, ..., v_m));$$
$$\Sigma' = \Sigma \cup \{\varphi^*, \psi^*\}.$$

The two new statements in Σ' are called *definitions* of the new function and relation.

A special case of a definition of a new function occurs when $m = 0$. In this case, if $\Sigma \vdash \exists! v_m \psi(v_m)$, then $\psi^{\mathfrak{A}}$ corresponds to a single element of A. In other words, ψ^* defines a new constant.

For example, if $\mathfrak{M} \models$ ZFC, and $\psi(x)$ is the formula "$\forall y\,(y \notin x)$", then $\psi^{\mathfrak{M}}$ is $\mathfrak{M}$'s version of the empty set. We shall write $\emptyset^{\mathfrak{M}}$ rather than $\psi^{\mathfrak{M}}$. Moreover, as we explained in Chapter 3, $\emptyset$ is the set-theoretic version of zero. Therefore, we shall sometimes write $0^{\mathfrak{M}}$ instead of $\emptyset^{\mathfrak{M}}$. Other natural numbers have similar definitions, and we shall in general write $n^{\mathfrak{M}}$ for $\mathfrak{M}$'s version of the natural number n.

EXERCISE 39(G): Find a formula ψ of L_S with one free variable such that $\psi^{\mathfrak{M}} = 1^{\mathfrak{M}}$ whenever $\mathfrak{M} \models$ ZFC.

Now let $\mathfrak{A} = \langle A, (R_i)_{i \in I}, (F_j)_{j \in J}, (C_k)_{k \in K} \rangle$ be a model of Σ. Denote

$$R = \varphi^{\mathfrak{A}}, \qquad F = \psi^{\mathfrak{A}};$$
$$\mathfrak{A}' = \langle A, (R_i)_{i \in I}, R, (F_j)_{j \in J}, F, (C_k)_{k \in K} \rangle.$$

Then

$$\mathfrak{A}' \models \Sigma'.$$

Moreover, let $\vec{x} = \langle x_0, ..., x_{k-1} \rangle$. To each formula $\varphi'(\vec{x})$ of L' there corresponds a formula $\varphi(\vec{x})$ of L such that for arbitrary $\vec{a} \in A^k$:

$$\mathfrak{A}' \models \varphi'[\vec{a}] \text{ iff } \mathfrak{A} \models \varphi[\vec{a}].$$

EXERCISE 40(PG): What we just described can be expressed by a theorem. Formulate such a theorem. Be especially careful to formulate all the assumptions of the theorem in their most general form.

EXERCISE 41(G): Let L'_S be the extension of L_S by the binary function symbols $\cup$ and $\cap$, and let $\varphi'(v_0, v_1)$ be the formula: "$\forall v_2\,(v_0 \cap v_1) \cup v_2 \subseteq v_0 \cup v_2$." Assuming that the meanings of $\cup$ and $\cap$ are defined in the standard way, find the formula $\varphi(v_0, v_1)$ of L_S that corresponds to $\varphi(v_0, v_1)$.

EXERCISE 42(R): Prove the theorem that you formulated in Exercise 40. *Hint:* Use induction over the wellfounded relation of being a subformula.

More often than not, we shall use expressions of plain English instead of new symbols to denote relations and functions on sets. Whenever we write $\mathfrak{A} \models$ "*FANCY PROPERTY*$(\vec{a})$," we really mean that $\mathfrak{A} \models \varphi(\vec{a})$, where φ is the formula of L that corresponds to our expression *FANCY PROPERTY* of the natural language. This is an important point to remember if one doesn't want to get carried away by all sorts of connotations of the natural language. Let us consider the case where $FANCY\ PROPERTY(x)$ is "x is a finite set." We must select some formula $fin(x)$ of L_S and agree among ourselves that

Definition: x is a finite set $\equiv$ $fin(x)$.

For example, $fin(x)$ could be: "$\exists n \in \omega\, (n \approx x)$." Having made a choice of fin, and having put down the above definition, we are free to use the expression "x is a finite set" whenever we feel like it. But we must guard against the illusion that if $\langle M, E\rangle \models$ ZFC, and $\langle M, E\rangle \models$ "x is a finite set," then x *is* really a finite set. First of all, if your metalanguage allows for objects that are not sets, then there is no reason why the elements of M should be sets in the first place. Second, even if the elements of your model are sets, the relation E does not have to be the membership relation on M. And third, even if $\langle M, E\rangle$ is a *standard model,* i.e., if xEy iff $x \in y$ for all x, y in M, it may still happen that all elements of M are infinite sets, although there are many x in M such that $M \models fin(x)$.

EXERCISE 43(PG): Suppose M is a family of sets, $\langle M, E\rangle$ is a model of L; and suppose X is an infinite set such that $X \cap TC(M) = \emptyset$. Let $M' = \{x \cup X : x \in M\}$, and $E' = \{\langle x \cup X, y \cup X\rangle : \langle x, y\rangle \in E\}$.

(a) Convince yourself that M' consists only of infinite sets, and that, nevertheless, for each sentence φ of L_S, $\langle M, E\rangle \models \varphi$ iff $\langle M', E'\rangle \models \varphi$.

(b) Show that if for some x in M the set $\{y \in M : \mathfrak{M} \models y \in x\}$ is finite, then so is the set $\{y' \in M' : \mathfrak{M}' \models y \in x \cup X\}$.

The following question naturally arises from the above exercises:

Question 20: Is it possible to find a formula $fin(x)$ of L_S such that whenever $\langle M, E\rangle \models$ ZFC and x is an element of M, then $\langle M, E\rangle \models fin(x)$ iff the set $\{y \in M : \langle M, E\rangle \models y \in x\}$ is finite?

We are going to show that the answer to Question 20 is "no." Let L_S^+ be the language one obtains by adding for each $k \in \omega$ a constant symbol c_k to L_S. Let $fin(x)$ be any formula of L_S that expresses a property shared by all finite sets. The expression "a property shared by all finite sets" is too vague for the present purpose; let us require that if n is a natural number, then ZFC $\vdash fin(n)$. Note that since L_S^+ extends L_S, $fin(x)$ is also a formula of L_S^+. Now consider the following theory Σ in L_S^+:

$$\begin{aligned}\Sigma = \text{ZFC } &\cup \{\text{“}c_k \neq c_\ell\text{”} : k, \ell \in \omega\backslash\{0\} \wedge k \neq \ell\}\\ &\cup \{\text{“}c_k \in c_0\text{”} : k \in \omega\backslash\{0\}\} \cup \{fin(c_0)\}.\end{aligned}$$

Let Σ^- be a finite subset of Σ, and let N be a positive natural number such that if a sentence "$c_k \neq c_\ell$" or "$c_k \in c_0$" is in Σ^-, then $k, \ell < N$. Assume that $\mathfrak{M} = \langle M, E\rangle$ is a model of ZFC. We can extend it to a model of L_S^+ by choosing an interpretation $C_k \in M$ for each of the constant symbols c_k. Given Σ^- and the corresponding N, we can do this by letting: $C_0 = N^{\mathfrak{M}}$; $C_k = k^{\mathfrak{M}}$ for $0 < k < N$; $C_k = N^{\mathfrak{M}}$ for $k \geq N$.

EXERCISE 44(PG): Show that if $\mathfrak{M}^+ = \langle M, E, (C_k)_{k\in\omega}\rangle$, then $\mathfrak{M}^+ \models \Sigma^-$ and $\mathfrak{M}^+ \not\models \Sigma$.

It follows that as long as ZFC has a model, so has every finite fragment of Σ. Thus by the Compactness Theorem 5.3, there exists a model

$$\mathfrak{M}^\bullet = \langle M^\bullet, E^\bullet, (C_k^\bullet)_{k\in\omega}\rangle$$

such that $\mathfrak{M}^\bullet \models \Sigma$. Fix such a model and let $\mathfrak{M}^\circ = \langle M^\bullet, E^\bullet\rangle$ be the model that one obtains by "forgetting" the meaning of the constants C_k.[24] Then $\mathfrak{M}^\circ$ is a model of L_S. The model $\mathfrak{M}^\circ$ is called the *L_S-reduct* of $\mathfrak{M}^\bullet$. Moreover, $\mathfrak{M}^\circ \models$ ZFC; since ZFC $\subset \Sigma$, and $\mathfrak{M}^\circ$ "remembers" all the information of $\mathfrak{M}^\bullet$ that is expressable in L_S. Now let x be the element of $M^\bullet$ which $\mathfrak{M}^\bullet$ recognizes as C_0. Since $fin(x)$ was supposed to be a formula of L_S, it is still true that $\mathfrak{M}^\circ \models fin(x)$. On the other hand, consider the set $Y = \{y \in M^\bullet : \mathfrak{M}^\circ \models y \in x\}$. This set contains all C_k's, and the C_k's are pairwise distinct.

EXERCISE 45(G): Why?

Of course, $\mathfrak{M}^\circ$ has "forgotten" whether or not a given y was a C_k, but it still "remembers" whether or not a given y is an element of x. Therefore, Y is infinite, and it follows that the answer to Question 20 is negative.

Note that in our whole discussion of Question 20 we did not need to know the axioms of ZFC. A choice of stronger axioms does not change the answer to Question 20. In other words, we have shown the following.

Theorem 21: *Finiteness of an element of a model is not a first-order property.*

Now let us return to the apparent paradox which was stated at the end of the previous section. It went like this. We started out with the observation that if $\langle M, E\rangle \models$ ZFC $+\neg CON$(ZFC), then there exists a p in M such that $\langle M, E\rangle \models$ "p is a proof of '$\emptyset = \{\emptyset\}$' from the axioms of ZFC." We tried to scare you a little by suggesting that one should be able to decode a real proof of "$\emptyset = \{\emptyset\}$" from p.

A proof is a finite sequence of formulas of L_S. Therefore, in particular, $\langle M, E\rangle \models fin(p)$. As we have shown above, this does *not* imply that

$$P = \{\varphi \in M : \langle M, E\rangle \models \varphi \in Form \wedge \varphi \text{ appears in the sequence } p\}$$

really must be a finite set. But, if P is infinite, how could one decode a real proof of "$\emptyset = \{\emptyset\}$" from p? We do not see how; and we hope you don't either.

We guess it's time to tell you what the axioms of ZFC are. This we shall do in the next chapter.

---

[24]Note that this does not mean deleting these elements from the universe of $\mathfrak{M}^\bullet$. The technique of building models by "forgetting" part of the information can be best illustrated by a real-life example. Suppose you are one of the people who couldn't care less about anniversaries and the like, but you are married and these things mean a lot to your spouse. Then you are like $\mathfrak{M}^\circ$, whereas your spouse is like $\mathfrak{M}^\bullet$. Both you and your spouse live in the same "universe of days." But your spouse's "language" contains a constant symbol "our wedding anniversary," whereas you "do not think in terms of anniversaries." Your spouse interprets one of the days of the year as "our wedding anniversary." You feel no urge to do the same, since your "language" does not contain the symbol "our wedding anniversary." Nevertheless, if your spouse interprets this day as your wedding anniversary, it probably is, and you may be in deep trouble.

Mathographical Remarks

The Completeness Theorem, the Compactness Theorem, and the concept of $\mathfrak{m}$-categoricity of a theory belong to a branch of mathematical logic called *model theory*. Model theorists keep themselves busy by studying methods for constructing models of formal theories, and also by classifying such models. The following books give an introduction to and overview of this area.

C. C. Chang and H. J. Keisler, *Model Theory*, 3rd ed., North-Holland, 1990.

W. Hodges, *Model Theory*, Cambridge University Press, 1993.

Hodges' book also has an introduction to second-order languages that is brief and to the point (pages 75–76).

Gödel announced his Incompleteness Theorems in 1930, and published the proof in his paper *Über formal unentscheidbare Sätze der Principia Mathematicae und verwandter Systeme. I*, Monatshefte für Mathematik und Physik 37, 349–360.

There are several excellent books from which you can learn more about Gödel's proof. The highly entertaining book *Gödel, Escher, Bach: An Eternal Golden Braid*, by D. Hofstadter, Basic Books, New York 1979, 777 pages, contains a slow-paced, thorough, nontechnical discussion of the two main ideas involved in the proof: coding and self-reference. Hofstadter shows parallels between Gödel's proof, Bach's music, and the drawings of Escher.

If you prefer a nontechnical introduction to the main ideas of the proof that is a little shorter than Hofstadter's, we recommend *Gödel's Proof*, by E. Nagel and J. R. Newman, New York University Press, 1958.

A well-written, rigorous presentation of several proofs of Gödel's Theorems as well as related results can be found in R. Smullyan, *Gödel's Incompleteness Theorems*, Oxford University Press, 1992, 139 pages.

The statements which we called the First and Second Foundational Theses do not have any special names elsewhere in the literature. Naming them in this text serves the sole purpose of illustrating the status of ZFC. The pompous names we chose should be taken with a grain of salt.

CHAPTER 7

The Axioms

In this chapter, we present the first-order theory ZFC. The letters stand for **Z**ermelo and **F**raenkel, who formulated the axioms, and for the Axiom of Choice. The axioms of ZFC are sentences of the language of set theory, L_S, whose only nonlogical symbol is the relational symbol $\in$ of arity 2.

Throughout our discussions of the axioms, you want to keep in mind their twofold nature: On the one hand, each axiom should express an obvious truth about naively understood sets and their membership relation. On the other hand, we are just presenting a first-order theory with no meaning *per se*. In particular, some subset of the theory, or even the whole theory ZFC, may have models of a very different nature from the usual interpretation. To stress the latter point, we shall show you some models for subtheories of ZFC. These models will be of two different types:

(++) *standard models* $\langle X, \overline{\in} \rangle$, where X is a set, and $\overline{\in}$ stands for the "real" (naively understood) membership relation on the sets in X. The distinction between $\overline{\in}$ and $\in$ will be made only in this chapter and in Chapter 12. It is supposed to help you see when $\in$ is treated as a relation, and when it is treated merely as a symbol.

(++) *nonstandard models* $\langle X, E \rangle$, where X is any nonempty set, and E is a binary relation on X other than $\overline{\in}$.

(A1) Axiom of Extensionality:

$$\forall x, y\, (x = y \leftrightarrow \forall z (z \in x \leftrightarrow z \in y)).$$

At the beginning of Chapter 1, we showed how this axiom follows from our naive notion of sets. Thus, extensionality nicely matches our intuition, at least as long as the variables x and y range only over sets. Let us mention that (A1) distinguishes ZFC from axiom systems for set theory which allow so-called *urelements*. Urelements are objects other than sets. They can be elements of sets, but do not themselves contain elements. For example, a real number like π is naively understood as an urelement. It has no elements, but it is a member of $\mathbb{R}$. Therefore, the structure $\langle \mathbb{R}, \overline{\in} \rangle$ does not satisfy (A1): Both π and $\sqrt{2}$ contain the same elements (none), but they are different objects.

Banishing urelements from our universe may seem the wrong thing to do, since most naively understood sets are ultimately built up from non-sets, that is, urelements. The approach taken here is a matter of convenience. Most arguments in set theory become clumsier if we have to consider urelements as well as sets. Take our axiom (A1). If we allow urelements, then the Axiom of Extensionality should say: "If x and y are sets, then $x = y$ if and only if x and y contain the same elements." To express this in a formal language, we would need some unary predicate symbol S so that $S(x)$ could be interpreted as "x is a set." Adding a symbol S to our language is not all that terrible, but what would we gain from the extra effort? Nothing. Every part of mathematics that can be formalized within set theory with urelements can also be formalized within ZFC. Thus, urelements fell victim to Ockham's razor. [1]

If there are no urelements, then sets can only be composed of other sets. According to the architect's view of sets, the process has to start somewhere. The only possible starting point is a set that requires no building blocks whatsoever—the empty set $\emptyset$.

(A2) Empty Set Axiom

$$\exists x \forall y \, (y \notin x).$$

By (A1), the empty set is unique.

EXERCISE 1(PG): (a) Show that $\langle \omega, \overline{\in} \rangle \models (A1) \wedge (A2)$.
(b) Show that $\langle \{\emptyset, \{\{\emptyset\}\}, \{\emptyset, \{\emptyset\}\}\}, \overline{\in} \rangle \models \neg(A1) \wedge (A2)$.

Note that if $\langle X, \overline{\in} \rangle$ is a standard model, and if $\mathbf{x}$ is an element of X such that $\langle X, \overline{\in} \rangle \models_s \forall y \, (y \notin x)[\mathbf{x}]$, then it is not necessarily true that $\mathbf{x} = \emptyset$. The set $\{\{\emptyset\}\}$ in Exercise 1(b) is an example. You may wonder why we did not simply write $\langle X, \overline{\in} \rangle \models \forall y \, (y \notin \mathbf{x})$ instead of $\langle X, \overline{\in} \rangle \models_s \forall y \, (y \notin x)[\mathbf{x}]$. If x were a constant symbol of our language, then the interpretation of this symbol would be a feature of a given model. But since x is a variable, its interpretation depends on the particular valuation s, and we have to write $\models_s$ instead of $\models$. However, if $\langle X, E \rangle$ is a model of L_S and $\mathbf{x}$ is the unique element of X such that $\langle X, E \rangle \models_s \forall y (y \notin x)[\mathbf{x}]$ for every valuation $s : \mathbb{N} \to X$, then we can use the convenient notation $\mathbf{x} = \emptyset^{\langle X, E \rangle}$. In particular, if $\langle X, E \rangle \models (A1) \wedge (A2)$, then a unique $\emptyset^{\langle X, E \rangle}$ exists.

The next exercise shows that $\langle X, \overline{\in} \rangle$ may satisfy the Empty Set Axiom even if X does not contain $\emptyset$.

EXERCISE 2(PG): (a) Show that $\langle \omega \backslash n, \overline{\in} \rangle \models (A1) \wedge (A2)$ for each $n \in \omega$.
(b) Given $n \in \omega$, find $\emptyset^{\langle \omega \backslash n, \overline{\in} \rangle}$.

Let us compare the models in Exercises 1 and 2. The standard model

$$\langle \{\emptyset, \{\{\emptyset\}\}, \{\emptyset, \{\emptyset\}\}\}, \overline{\in} \rangle$$

does not satisfy the Axiom of Extensionality because $X = \{\emptyset, \{\{\emptyset\}\}, \{\emptyset, \{\emptyset\}\}\}$ is not transitive. The only element of $\{\{\emptyset\}\}$, the set $\{\emptyset\}$, is not in X. Therefore, $\langle \{\emptyset, \{\{\emptyset\}\}, \{\emptyset, \{\emptyset\}\}\}, \overline{\in} \rangle \models \forall z \, (z \in \{\{\emptyset\}\} \leftrightarrow z \in \emptyset)$. On the other hand, by Exercise 4.4(c), the set ω is transitive. As you proved in your solution to Exercise 1(a), $\langle \omega, \overline{\in} \rangle \models (A1)$. Let us now generalize your argument.

Theorem 1: *If X is a transitive set, then $\langle X, \overline{\in} \rangle \models (A1)$.*

Proof: Let X be a transitive set. We want to show that

$$\langle X, \overline{\in} \rangle \models \forall v_0, v_1 \, (v_0 = v_1 \leftrightarrow \forall v_2 \, (v_2 \in v_0 \leftrightarrow v_2 \in v_1))$$

---

[1] "Ockham's razor" is the principle that in building a scientific or philosophical theory one should not postulate the existence of unnecessary entities.

In view of the definition of the symbol $\models$, this means that for every valuation $s : \mathbb{N} \to X$, the equality $s(0) = s(1)$ holds if and only if for all $s^* : \mathbb{N} \to X$, the equality $s^*(0) = s^*(1)$ implies that ($s^*(2)\overline{\in}s^*(0)$ iff $s^*(2)\overline{\in}s^*(1)$).

EXERCISE 3(PG): Recall the definition of $\models$ from Chapter 5. Convince yourself that we correctly translated the statement "$\langle X, \overline{\in}\rangle \models$ (A1)" into the terminology of valuations.

Now observe that since we care only about $s(0), s(1)$, and $s(2)$, quantifying over entire valuations amounts to carrying an excessive amount of unnecessary luggage. When we defined the satisfaction relation $\models_s$ in Chapter 5, it was convenient to quantify over arbitrary valuations. This allowed us to minimize the number of symbols used in the definition. But whenever you analyze the relation $\models_s \varphi$ for a specific formula φ, you will need to consider only the values $s(n)$ for the finitely many n for which v_n actually appears in φ (see Exercise 5.5 for a variation on this theme). Therefore, you may want to translate the symbol $\mathfrak{M} \models_s \varphi$ (or $\mathfrak{M} \models \varphi$) into a statement that involves quantification over elements of the universe of $\mathfrak{M}$ instead of quantification over valuations. Here is such a translation for "$\langle X, \overline{\in}\rangle \models$ (A1)":

(+) For all x_0, x_1 in X, x_0 is equal to x_1 if and only if for all x_2 in X ($x_2\overline{\in}x_0$ iff $x_2\overline{\in}x_1$).

In our translations, we have been very careful about the style of symbols that we used for logical connectives and the membership relation. Symbols like $\forall, \leftrightarrow$ and $\in$ were used only for the language of the model. Plain English was used for the metalanguage. Moreover, we used $\overline{\in}$. For good reason.

EXERCISE 4(G): Convince yourself that if you replace in (+) all occurences of $\overline{\in}$ by E, then you get a translation of $\langle X, E\rangle \models$ (A1) for an arbitrary (possibly nonstandard) model $\langle X, E\rangle$ of L_S.

Now we prove (+). Let X be a transitive set, and let $x_0, x_1 \in X$. If x_0 is equal to x_1, then x_0 and x_1 have the same elements. Therefore, for all x_2 in X, $x_2\overline{\in}x_0$ iff $x_2\overline{\in}x_1$.

For the other implication, assume x_0 is not equal to x_1. Then there exists some x_2 which is in the symmetric difference of x_0 and x_1. To be specific, assume x_2 is in x_0 but not in x_1 (the other case is symmetric). Since x_0 is in X and X is transitive, x_2 is also an element of X. Thus there exists an x_2 in X such that the biconditional "$x_2\overline{\in}x_0$ iff $x_2\overline{\in}x_1$" fails. This proves the "if"-direction of (+). The proof of Theorem 1 is complete. □

Exercise 2(a) shows that transitivity is not a necessary condition for $\langle X, \overline{\in}\rangle$ to satisfy the Axiom of Extensionality: For $n > 0$, the set $\omega\backslash n$ does not contain $\emptyset$, but by Exercise 4.4(b), $\emptyset \in TC(n+1) \subseteq TC(\omega\backslash n)$. Therefore, the set $\omega\backslash n$ is not transitive. Nevertheless, $\langle \omega\backslash n, \overline{\in}\rangle \models$ (A1).

Now let $E = \{\langle n+1, n\rangle : n \in \omega\}$. Then $\langle \omega, E\rangle$ is a nonstandard model of L_S.

EXERCISE 5(PG): Show that $\langle \omega, E\rangle \models \neg$(A2).

Can the Empty Set Axiom also fail in a standard model $\langle X, \overline{\in}\rangle$? No. In Chapter 4 we mentioned that the relation $\overline{\in}$ is strictly wellfounded.

EXERCISE 6(PG): (a) Show that if E is a strictly wellfounded binary relation on a set X, then $\langle X, E\rangle \models$ (A2).

(b) Conclude that the Empty Set Axiom holds in every standard model.

Most of the remaining axioms will specify what procedures can be used to construct new sets from given ones. This should not come as a surprise. Recall that Russell's Paradox arises from an attempt to collect too many sets into another set. We are trying to eliminate this paradox by adopting the architect's view that sets can only be assembled from ready-made building blocks (i.e., given sets) according to sound construction methods. The remaining axioms (except the Axiom of Infinity and the Foundation Axiom) describe sound construction methods.

EXERCISE 7(G): What is the simplest possible way of forming a new set from given ones? Try to formulate an axiom that describes it.

To help you in resisting the temptation to read on before doing Exercise 7, we delay the answer and first discuss another axiom.

(A3) Foundation Axiom:

$$\forall x(x \neq \emptyset \rightarrow \exists y \in x \forall z \in x(z \notin y)).$$

This axiom expresses the fact that $\overline{\in}$ is strictly wellfounded. As you already know, wellfoundedness of $\overline{\in}$ implies that no set x contains itself as an element, that there are no finite sequences $\langle x_0, x_1, \dots, x_k\rangle$ with $x_0 \overline{\in} x_1 \overline{\in} \dots \overline{\in} x_k \overline{\in} x_0$, and no infinite sequences $\langle x_n : n \in \omega\rangle$ such that $x_{n+1} \overline{\in} x_n$ for all $n \in \omega$. If you paid close attention to our distinction between $\in$ and $\overline{\in}$, you are probably puzzled why we used $\overline{\in}$ instead of $\in$. Of course, the axiom should express some truth about $\overline{\in}$. This it does. Does it also imply wellfoundedness of every relation E on a set X such that $\langle X, E\rangle \models$ (A3)?

EXERCISE 8(PG): Let $\langle X, E\rangle$ be the model of Exercise 5. Show that E is not wellfounded, and yet $\langle X, E\rangle \models$ (A3).

How come? The problem may seem to arise from the fact that $\langle X, E\rangle$ is not a model of all the axioms of ZFC, but this is not the crux of the matter. In Chapter 13 we shall show how it is possible to construct models $\langle Y, F\rangle$ for all of ZFC so that F is not a wellfounded relation. This phenomenon can occur because wellfoundedness is not a first-order property. In Exercise 6.27, you were asked to convince yourself that the statement $\langle Y, F\rangle \models$ "$\in$ is wellfounded" does not necessarily imply that F is a wellfounded relation on Y. Let us now see how this phenomenon arises in the special case of the model $\langle x, E\rangle$ of Exercise 5.

While you, looking at $\langle X, E\rangle$ from the outside, see that the universe X of this model is not a wellfounded set, the model does not "see" this. The model "sees" only the elements of its underlying set, and for each x that it "sees," it also "knows"[2] of some yEx such that $\neg zEy$ holds for all zEx (if $x = n$, then the y is $n+1$. Note

[2]This anthropomorphic language is not one of our idiosyncrasies. Many logicians speak (but less frequently write) of models that see, know, or believe.

that the unique z such that zEy is $n+2$, and this z does not satisfy zEx).

Let us contrast the second-order property of strict wellfoundedness with a first-order consequence of it.

The statement $\varphi : \forall x_0, x_1 \neg(x_0 \in x_1 \wedge x_1 \in x_0)$ is a theorem of ZFC. To see this, suppose φ is false, and let x_0, x_1 be a counterexample. Form the set $x = \{x_0, x_1\}$.

EXERCISE 9(G): Show that this set x witnesses the negation of (A3).

To make the above argument work, we need an axiom that allows us to construct the set $\{x_0, x_1\}$ from the building blocks x_0 and x_1.

(A4) Pairing Axiom:

$$\forall x, y \exists z \forall u (u \in z \leftrightarrow u = x \vee u = y).$$

In other words, the Pairing Axiom allows us to form unordered pairs $\{x, y\}$. Did you get this axiom as your solution of Exercise 7? The only even simpler one that we (the authors) can think of would be a "Singleton Axiom":

$$\forall x \exists y \forall z\, (z \in y \leftrightarrow z = x).$$

But since $\{x\} = \{x, x\}$, the Singleton Axiom follows from the Pairing Axiom and is therefore redundant.

EXERCISE 10(PG): Let $X = \{\emptyset\}$ and $E = \{\langle \emptyset, \emptyset \rangle\}$. Show that $\langle X, E \rangle \models$ (A4).

By applying the Pairing Axiom twice, we can form an ordered pair $\langle x, y \rangle = \{\{x\}, \{x, y\}\}$. This is important, since ordered pairs are the building blocks of binary relations, and, in particular, of functions. Can we also form unordered triples $\{x, y, z\}$ by repeated application of the Pairing Axiom alone? Try. You will not succeed. How can we know? Of course, *we* did not succeed, but this is no proof that nobody else might be lucky. In Chapter 6 we explained that in order to prove that a sentence φ is not a theorem of a theory T, it suffices to construct a model of T where the negation of φ holds. Let us show how this can be done for the theory $\{(\text{A4})\}$ and the sentence φ:

$$\forall x, y, z \exists w \forall u\, (u \in w \leftrightarrow u = x \vee u = y \vee u = z).$$

EXERCISE 11(PG): (a) Suppose $|X| \geq 2$ and $\langle X, \overline{\in} \rangle \models$ (A4). Let $Y = \{x \overline{\in} X : |x| \leq 2\}$. Prove that $\langle Y, \overline{\in} \rangle \models$(A4)$\wedge\neg\varphi$.

(b) Formulate and prove the analogue of (a) for nonstandard models $\langle X, E \rangle$.

For the formation of unordered triples, quadruples, etc., one more axiom is needed.

(A5) Union Axiom:

$$\forall x \exists y \forall z (z \in y \leftrightarrow \exists u (u \in x \wedge z \in u)).$$

As you already know, we write $\bigcup x$ for the union of x, that is, $\bigcup x = \{z : \exists u \in x (z \in u)\}$.

EXERCISE 12(G): Let a, b, c be arbitrary sets.
(a) Show that $a \cup b = \bigcup\{a, b\}$.
(b) Show that $\{a, b, c\} = \bigcup\{\{a, b\}, \{b, c\}\}$.

You may now expect that the next axiom will be an "intersection axiom," followed perhaps by a "difference axiom." In a sense, you are right. But the following construction principle is much more general than that.

(A6) Comprehension Axiom Schema: [3] For each formula $\varphi(x)$ of L_S which does not contain a free occurrence of the variable y, the following is an axiom:

(A6φ) $$\forall z \exists y \forall x (x \in y \leftrightarrow x \in z \wedge \varphi(x)).$$

We can describe (A6) as follows: Let z be a set, and let $\mathcal{E}$ be a property of sets that can be expressed in the language of set theory by a formula $\varphi(x)$. Then all elements of z that have property $\mathcal{E}$ form a set. We denote the set y which consists of all x in z that satisfy $\varphi(x)$ by $\{x \in z : \varphi(x)\}$. In a first-order language, we are allowed to quantify over variables, but not over such objects as formulas. An expression like $\forall \varphi$ cannot be formed in L_S. Thus, the first part of (A6) is an expression in the *metalanguage,* i.e., in the language we use to talk *about* L_S. Therefore, it is formulated in plain English, whereas (A6φ) is an expression of L_S, and can be rendered symbolically. We shall refer to the axiom (A6φ) as an *instance* of the axiom schema (A6).

It is not possible to formulate a single sentence of L_S that would be equivalent to the conjuction of all instances of (A6). But this is no problem. Recall that if A is any set of sentences of a language L such that there is a purely mechanical procedure for deciding whether a given sentence of L is in A, then A can serve as a set of axioms.

EXERCISE 13(PG): How might such a decision procedure for (A6) look?

Let us now consider some instances of (A6). Denote by $\varphi_0(x)$ the formula $x \neq x$. Let z be any set. By (A6φ_0), the set $z^* = \{x \in z : \varphi_0(x)\}$ exists. This set does not contain any elements; it is the empty set. Hence, if there exists any set z whatsoever, then the empty set exists. One can therefore replace the Empty Set Axiom by the *Set Existence Axiom,* which states that there is at least one set. The resulting theory is equivalent to our version of ZFC.

Now consider the formula $\varphi_1(x)$: $x \notin x$. When we tried to form the set $Y = \{x : x \notin x\}$, we got Russell's Paradox. But, for any given set z, Axiom (A6φ_1) entitles us to form the set $y = \{x \in z : x \notin x\}$. By the Foundation Axiom, $y = z$. But, since $z \notin z$, the question whether or not $y \in y$ will never have to be considered in building the set y; and we get no paradox.

Finally, consider $\varphi_2(x)$: $x \in q$. This formula reveals another important aspect of (A6). The formulas φ that can generate instances of it may contain *parameters.* In our first application of (A6φ_2), x will be a free variable, and q will be a parameter. The distinction between free variables and parameters can be confusing

[3] The Comprehension Axiom Schema is frequently called *Separation Axiom Schema,* or *Separation Axiom.*

for the beginner. No intrinsic property of the expression "$x \in q$" makes q more "parametrish" than x. Either of x, q could *act* as a free variable or a parameter, depending on the context in which we use the formula. This is similar to many familiar distinctions. Is farmer Brown a consumer, or is he a producer? This depends on which commodity we are interested in. He is a producer of wheat and a consumer of toothpaste; take your pick. The distinction between free variables and parameters is best understood if you look at many examples and get used to the terminology.

Now let q, z be sets. By (A6φ_2), the set $\{x \in z : x \in q\} = z \cap q$ is a set. A reversal of the roles of parameter and free variable implies that if x, z are sets, then so is $y = \{q \in z : x \in q\}$. For example, if $z = \omega$, and $n \in \omega$, then $y = \{q \in z : n \in q\} = \{m \in \omega : m > n\}$.

EXERCISE 14(G): Formulate an instance (A6φ_3) of the Comprehension Axiom Schema which implies that for every two sets z, q the difference $y = z \backslash q$ exists.

There is one more Axiom Schema in ZFC:

(A7) Replacement Axiom Schema: For each formula $\varphi(x)$ of L_S which does not contain a free occurrence of the variable y, the following is an axiom:

(A7φ) $$\forall a\, (\forall x \in a \exists! y\, \varphi(x, y) \rightarrow \exists z \forall x \in a \exists y \in z\, \varphi(x, y)).$$

Frequently, this axiom schema is somewhat sloppily called the *Replacement Axiom*. Apparently, this Axiom Schema assures us the range of every definable function f is a set. But do we need a new axiom schema for this?

EXERCISE 15(PG): (a) Show that if f is a function with domain a, then

$$rng(f) = \{y \in \bigcup \bigcup f : \exists x \in a\, (\{\{x\}, \{x, y\}\} \in f)\}$$

(b) Conclude that if f is a function, then it follows from (A4), (A5) and (A6) that $rng(f)$ is also a set.

So, what is (A7) good for? Put yourself in the shoes of the architect of the set-theoretical universe. Suppose you have already constructed a set a, and you know a formula φ such that for every $x \in a$, there is exactly one y such that $\varphi(x, y)$ holds. Does this also mean that you have already constructed a function f with domain a such that $\varphi(x, f(x))$ holds for all $x \in a$? Not necessarily. You may try to use an instance of (A6) for the construction of such a function. So, you take a set b with the property that for each $x \in a$ there exists $y \in b$ such that $\varphi(x, y)$, then you form $a \times b$, and finally, you let $f = \{\langle x, y \rangle : \varphi(x, y)\}$. Sounds easy. A little too easy, as a matter of fact. First of all, where do you take the set b from? For each $x \in a$, you have already constructed the *individual* set y such that $\varphi(x, y)$ holds. But there is no reason why you should already have constructed a set b that would be big enough to contain all the y's needed. Axiom (A7φ) helps you out in this situation. It assures you that there is a set z which contains all the y's that you need for your construction of f. The proof of Theorem 3 given below contains a typical example of an application of (A7). In Chapter 12, we shall construct a model for all the axioms of ZFC *except* the Replacement Axiom Schema.

So far, so good. But the "easy" construction of f in the last paragraph contains one more gap.

EXERCISE 16(G): Can you detect it?

Of course, in the "easy" construction of f we said "then you form $a \times b$," without justifying this construction on the grounds of axioms that had been described up to that point. The problem is not solved by the use we made of (A7). We have a set z that contains all relevant y's, but in order to construct the function f itself, we still must form $a \times z$, and then use (A6) to "separate out" the pairs that form f. The elements of $a \times z$ are ordered pairs of previously constructed sets, and each such pair individually can be formed by applying the Pairing Axiom twice. But how to collect all these pairs into one set? We need another axiom.

(A8) Power Set Axiom:

$$\forall x \exists y \forall z (z \in y \leftrightarrow z \subseteq x).$$

As you know, the set of all subsets of a set x is called the *power set of* x and denoted by $\mathcal{P}(x)$.

Let a, b be two sets, and assume $x \in a, y \in b$. Then $\{x\}, \{x, y\} \in \mathcal{P}(a \cup b)$, and $\langle x, y \rangle = \{\{x\}, \{x, y\}\} \in \mathcal{P}(\mathcal{P}(a \cup b))$. Hence, $a \times b \subseteq \mathcal{P}(\mathcal{P}(a \cup b))$. The right hand side of this inclusion contains more sets than the left hand side. But the Comprehension Axiom Schema allows us to form the set on the right hand side.

EXERCISE 17(PG): Find a formula $\varphi(z)$ such that

$$a \times b = \{z \in \mathcal{P}(\mathcal{P}(a \cup b)) : \varphi(z)\}.$$

The Power Set Axiom is a construction principle that would make an architect perfectly happy, but would drive an engineer nuts.[4] When designing a building, an architect must know what is technically feasible, but does not need to know all the details of the construction. In effect, the Power Set Axiom tells the architect of a set-theoretic universe: "Whenever you feel like collecting all subsets of a given set into a new one, go ahead." Great. But the engineer who is supposed to design the actual collecting procedure will not be so easily satisfied. How can she make sure that no subset is being missed in the process? The only axioms that tell us which sets *must* be subsets of a given set are the instances of (A6). But are all subsets of a given set z definable by some formula?

EXERCISE 18(PG): Show that if z is a finite set, and if $y \subseteq z$, then there exists a formula $\varphi(x)$ of L_S (possibly with parameters) such that $y = \{x \in z : \varphi(x)\}$.

EXERCISE 19(R): Let Y be the collection of all $y \subseteq \omega$ for which there exists a formula $\varphi(x)$ *without parameters* such that $y = \{x \in \omega : \varphi(x)\}$. Show that Y is countable.

Since $\mathcal{P}(\omega)$ is uncountable, Exercise 19 implies that there must be some entirely "wild" subsets of ω, or at least some subsets of ω that cannot be defined without

[4]Such axioms are called *nonconstructive axioms* by less irreverent authors. The Axiom of Choice is the other nonconstructive axiom.

the help of parameters. Now, our engineer has to make a tough decision. One of her options is to consider only those subsets of z that are definable by some formula, possibly with parameters. If she takes this approach, she will build the *constructible universe* **L**, which will be discussed in Chapter 12. Other options are less austere and allow the inclusion of some "wild" sets. The next exercise illustrates one way of producing a "wild" subset y of ω. When doing it, take an occasional break to relax and read the rest of this book.

EXERCISE 20(X): Flip a coin ω times. If it comes up heads on the n-th toss, put n into y. If it comes up tails on the n-th toss, don't put n into y. There is no reason why the resulting set y should be definable; even with parameters. With probability one, it won't be.

To sum up the previous paragraph: The axioms of ZFC do not impose strict limitations on the creation of subsets of a given infinite set. This allows set theorists to create various models of set theory. You will perhaps not be too surprised if we tell you that the Continuum Hypothesis holds in the austere model **L**, whereas the inclusion of "wild" subsets of ω can result in models where CH fails badly.

Now we show how (A7) and (A8) can be used to strengthen Theorem 4.25. Let us begin with an example.

Definition 2: Let x be any set, and let n be a natural number. Define $S^{(n)}(x)$, *the n-th successor of x,* recursively by:

$$S^{(0)}(x) = x; \qquad S^{(n+1)}(x) = S^{(n)}(x) \cup \{S^{(n)}(x)\}.$$

We have already seen a special case of this definition: The numeral $\tilde{n}$ is nothing else but $S^{(n)}(\emptyset)$.

For every $n \in \omega$, it is possible to find a formula $\varphi_n(x,y)$ of L_S such that $\varphi_n(x,y)$ holds iff $y = S^{(n)}(x)$. It is not hard to see that $\varphi_0(x,y)$: "$x = y$" and $\varphi_1(x,y)$: "$\forall z\,(z \in y \leftrightarrow (z \in x \vee z = x))$" are such formulas.

EXERCISE 21(G): (a) Find a formula $\varphi_2(x,y)$ such that $\varphi_2(x,y) \leftrightarrow y = S^{(2)}(x)$.
(b) Find a formula $\varphi_3(x,y)$ such that $\varphi_3(x,y) \leftrightarrow y = S^{(3)}(x)$.

The existence of $\varphi_{27}(x,y)$ assures us that an expression like "$\exists y\,(y = S^{(27)}(\omega))$" is indeed an abbreviation of a formula of L_S. However, as you saw in Exercise 21, the length of the formulas $\varphi_n(x,y)$ keeps growing with increasing n. Therefore, none of the formulas $\varphi_n(x,y)$ would be suitable for transcribing an expression like "$\forall n \in \omega \exists y\,(y = S^{(n)}(\omega))$" into L_S. The question arises whether there is a single formula $\psi(x,y,n)$ of L_S such that for all sets x,y and all $n \in \omega$, the formula $\psi(x,y,n)$ holds iff $y = S^{(n)}(x)$. To answer this question, we need the following modification of Theorem 4.25.

Theorem 3 (Definability by Recursion over a Wellfounded Relation): *Let $Z \neq \emptyset$, $k \in \omega\backslash\{0,1\}$, and let $a_3, \dots, a_k$ be arbitrary sets. Suppose that $\varphi_W(v_0, \dots, v_k)$ and $\varphi_G(v_0, \dots, v_k)$ are formulas of L_S all of whose free variables are among $v_0, \dots, v_k$ such that:*

- $W = \{\langle y, z\rangle : \varphi_W(y, z, a_3, \dots, a_k)\}$ *is a wellfounded relation on* Z*;*
- *for every function* f *whose domain is of the form* $I_W(z)$ *for some* $z \in Z$*, there exists exactly one* x *such that* $\varphi_G(f, z, x, a_3, \dots, a_k)$ *holds.*

Then there exists a formula $\varphi_F(v_0, \dots, v_k)$ *of* L_S *all of whose free variables are among* $v_0, \dots v_k$ *such that* $\forall z \in Z \exists! x\, \varphi_F(z, x, a_3, \dots, a_k)$*, and*

$$\text{(R)} \qquad \varphi_F(z, x, a_3, \dots, a_k) \leftrightarrow \varphi_G(F|I_W(z), z, x, a_3, \dots, a_k),$$

for each $z \in Z$ *(where* $F|I_W(z) = \{\langle y, x\rangle : \langle y, z\rangle \in W \wedge \varphi_F(y, x, a_3, \dots, a_k)\}$*).*

Moreover, if $\tilde{\varphi}_F(v_0, \dots v_k)$ *is another formula such that* (R) *holds, then*

$$\forall z \in Z \forall x\, (\tilde{\varphi}_F(z, x, a_3, \dots a_k) \leftrightarrow \varphi_F(z, x, a_3, \dots a_k)).$$

The formulas φ_W, φ_G, and φ_F are definitions of W, G, and F of Theorem 4.25. The sets $a_3, \dots, a_k$ are parameters of these definitions. Note that we do not require that all parameters are actually used in all definitions. If $k = 2$, then we have *definitions without parameters.*

Theorem 4.25 is a special case of Theorem 3. To see this, let W and G be as in the assumptions of Theorem 4.25. Let $k = 4$, $a_3 = W$, and $a_4 = G$. Consider the formulas: φ_W: $\langle v_0, v_1\rangle \in v_3$; φ_G: $\langle v_0, v_1, v_2\rangle \in v_4$.

Theorem 4.25 differs from Theorem 3 in two important respects: While the former theorem just proclaims the existence of a function F that satisfies (R), the latter also tells us that whenever W and G can be defined from parameters $a_3, \dots a_k$ (or without any parameters if $k = 2$), then the same is true for the function F. Moreover, Theorem 4.25 can be applied only if we are already given a set X which contains all possible function values of F. In practice, it may be difficult to find such a set X before the actual construction of F. Theorem 3 relieves us of the necessity to do so.

Here is a typical application of Theorem 3:

Corollary 4: *There exists a formula* $\varphi_F(v_0, v_1, v_3, v_4)$ *of* L_S *with four free variables such that for every* $n \in \omega$ *and arbitrary sets* x, y *the formula* $\varphi_F(n, y, x, \omega)$ *holds if and only if* $y = S^{(n)}(x)$.[5]

Proof: Let $Z = \omega$ and consider the formulas φ_W: $v_0 \in v_1$ and

$$\begin{aligned}\varphi_G : &\, fnc(v_0) \wedge dom(v_0) \in v_4 \wedge (v_0 = \emptyset \to v_2 = v_3) \\ &\wedge (\forall v_8, v_9\, (v_0 = v_8 \cup \{v_8\} \wedge \langle v_8, v_9\rangle \in v_0) \to v_2 = v_9 \cup \{v_9\}),\end{aligned}$$

where "$fnc(v_0)$" is an abbreviation of

$$\begin{aligned}``&\forall v_5 \in v_0 \exists v_6, v_7\, (v_5 = \langle v_6, v_7\rangle) \\ &\wedge \forall v_5, v_6, v_7\, (\langle v_5, v_6\rangle \in v_0 \wedge \langle v_5, v_7\rangle \in v_0 \to v_6 = v_7)."\end{aligned}$$

The following exercise concludes the proof of Corollary 4.

EXERCISE 22(G): Convince yourself that the assumptions of Theorem 3 are satisfied. Moreover, convince yourself that for $n \in \omega$ and arbitrary sets x, y, the formula $\varphi_F(n, y, x, \omega)$ given by Theorem 3 holds if and only if $y = S^{(n)}(x)$. □

---

[5]We shall see in Chapter 10 how we can get rid of the parameter ω in this corollary.

Note that in the proof of Corollary 4, we did not have to produce up front a set X such that $S^{(n)}(x) \in X$ for all $n \in \omega$.

EXERCISE 23(G): Use an instance of (A7) to show that for each set x there exists a set X such that $X = \{S^{(n)}(x) : n \in \omega\}$.

EXERCISE 24(X): Without using an instance of (A7), construct a set X such that $S^{(n)}(\omega) \in X$ for all $n \in \omega$.[6]

One can apply Theorem 3 to iterations of operations other than $S(x)$. In Definition 4.4, we introduced the symbol $\bigcup^{(n)} x$ for a natural number n and an arbitrary set x.

EXERCISE 25(PG): (a) Show that there exists a formula φ of L_S such that for all $n \in \omega$ and arbitrary sets x, the formula $\varphi(n, y, x, \omega)$ holds if and only if $y = \bigcup^{(n)} x$.

(b) Conclude that if ω is a set, then for every set x the transitive closure $TC(x)$ is again a set.

Proof of Theorem 3: Let Z, φ_W, φ_G be as in the assumptions, and let $\varphi_F(z, x, a_3, \dots, a_k)$ be the following formula: "$z \in Z \wedge \forall f$ (if f is a function with $I_W(z) \subseteq dom(f) \subseteq Z$ such that (a) $\forall z' \in dom(f)\, I_W(z') \subseteq dom(f)$, and (b) $\varphi_G(f|I_W(z'), z', f(z'), a_3, \dots, a_k)$ holds for all $z' \in dom(f)$, then

$$\varphi_G(f|I_W(z), z, x, a_3, \dots, a_k)))."$$

The following exercise concludes the proof of Theorem 3.

EXERCISE 26(R): (a) The formula φ_F has been rendered in a rather informal version of L_S. Convince yourself that it really can be expressed in L_S.

(b) Prove that φ_F satisfies the conclusion of Theorem 3.

(c) Show that the formula $\tilde{\varphi}_F$ which results from replacing the quantifier "$\forall f$" in the definition of φ_F by "$\exists f$" and changing the implication following this quantifier into a conjunction also satisfies the conclusion of Theorem 3. □

As in the proof of Theorem 4.25, the main goal of the above argument was avoiding self-reference in the definition of the function F. We achieved this goal by quantifying over all functions f with a certain property. Although we ostentatiously defined a function F by pointing to (all) other functions with a certain property, it so happens that the only function with the desired property is a suitable restriction of F. But this fact can be proved *after* giving the definition of F, and is not germane to the formulation of the definition itself.

EXERCISE 27(PG): Convince yourself that the proof of Theorem 4.25 uses essentially the same trick by pointing out the place(s) in the proof where the quantifier is hidden.

---

[6]Here and in Exercise 23, you may assume that ω is a set.

(A9) Axiom of Choice: For every family x of nonempty, pairwise disjoint sets, there exists a set z such that $|z \cap y| = 1$ for each $y \in x$.

EXERCISE 28(G): Write (A9) as a sentence of L_S. The only abbreviations you are allowed to use in this exercise are $\notin, \neq$ and $\emptyset$.

The Axiom of Choice generalizes the familiar process of choosing representatives. If x is the family of sets of congressional districts,[7] then the Congress of the United States is an example of a set z as in (A9). Note that congressional districts are nonempty and disjoint.

What makes the Axiom of Choice so special that we even mention it specifically in the name of our theory **ZFC**? Its nonconstructive character. The axiom assures us that there will be a Congress after the next Election Day, but it does not specify who will be elected. Otherwise, there would be no point in holding elections in the first place. The latter option has always been preferred by dictators, who evidently do not have much confidence in the Axiom of Choice and prefer legislatures composed of loyal supporters.

Now suppose for the sake of argument that you seized dictatorial power of the United Sets of the Universe. You need to stay on good terms with your archrival, the Former Solid Intersection, who keeps meddling in your internal affairs with evil propaganda about human rights, democracy and similar nuisances. In order to keep up appearances, you need some sort of congress. So you start looking at the list of congressional districts and make *your* choices. But there is a problem: The United Sets of the Universe have infinitely many congressional districts. While you are busy picking your best congress, your enemies may be plotting your overturn. You better get quickly done with this congress-choosing business. (By the way, how is Exercise 20 going? You must be taking a break from it, otherwise you would not be reading this.) In other words, you must make infinitely many choices in a finite amount of time without having the option of describing all these choices by a single formula. The Axiom of Choice says that you can somehow do this, but it does not give you much control over the outcome of the selection process.

EXERCISE 29(G): Did you ever before encounter the problem of having to make infinitely many choices without having precise instructions for which elements to choose? If so, in which context?

Right! We encountered exactly the same problem when we discussed recursive constructions. Theorem 4.26 has the same nonconstructive flavor as the Axiom of Choice. Moreover, the Axiom of Choice allows us to prove Theorem 4.26. For your convenience, we reproduce the theorem here before proving it.

Theorem 5 (Principle of Generalized Recursive Constructions): *Let $X, Z \neq \emptyset$, and let W be a wellfounded relation on Z. Moreover, suppose that G^* is a function with values in $\mathcal{P}(X)\backslash\{\emptyset\}$ such that $dom(G^*)$ consists of all pairs of the form $\langle f, z\rangle$ where $z \in Z$ and f is a function that maps $I_W(z)$ into X.*

Then there exists a function $F : Z \to X$ such that

$$(\mathrm{R}^*) \qquad F(z) \in G^*(F|I(z)) \quad \textit{for each } z \in Z.$$

[7]We understand a congressional district as a set of citizens eligible to be elected to congress.

Proof: Let X, Z, W, and G^* be as in the assumptions. The family $\mathcal{G} = \{G^*|\{\langle f, z\rangle\} : \langle f, z\rangle \in dom(G^*)\}$ is a family of nonempty, pairwise disjoint sets. By (A9), there exists a set G such that $|G \cap y| = 1$ for each $y \in \mathcal{G}$. Fix such G.

EXERCISE 30(G): (a) Show that G as chosen above, together with X, Z, and W, satisfies the assumptions of Theorem 4.25.

(b) Now apply Theorem 4.25 to derive the conclusion of Theorem 5. □

EXERCISE 31(X): Theorem 5 shares with Theorem 4.25 the unpleasant requirement that a set X which contains all possible values for F should be given before the actual construction of F. Formulate and prove a version of Theorem 5 that does not have a reference to such a set X in its assumptions.

Have we described all the tools needed to do set theory? No. Remember that a set x is hereditarily finite (see Chapter 4) if $TC(x)$ is finite. In Chapter 12 we shall see that the collection of hereditarily finite sets forms a set. This set will be denoted by V_ω.

EXERCISE 32(PG): Convince yourself that the standard model $\langle V_\omega, \overline{\in}\rangle$ satisfies all the axioms discussed so far in this chapter.

We need one more axiom to make the leap from the finite to the infinite.

(A10) Axiom of Infinity:

$$\exists x(x \neq \emptyset \wedge \forall u(u \in x \rightarrow u \cup \{u\} \in x)).$$

Let us show that this axiom indeed implies the existence of an infinite set. Let x be a nonempty set such that $u \cup \{u\} \in x$ for each $u \in x$. Let $u \in x$. Then also $u \cup \{u\}$, $u \cup \{u\} \cup \{u \cup \{u\}\}, \ldots$ are elements of x, and it follows from the previous axioms that these elements are pairwise distinct. Therefore, x must contain infinitely many elements.

Since the axioms (A1)–(A9) do not imply the existence of infinite sets, in particular they do not imply that ω is a set. For this reason we used strange phrases like "assume that ω is a set" earlier in this chapter. Axiom (A10) makes further repetition of such phrases unnecessary.

EXERCISE 33(PG): Prove from the axioms (A1)–(A10) that ω is a set.

That's it. ZFC is the collection of all axioms in (A1)–(A10). This collection is infinite, but recursive. It contains all the certainties on which contemporary mathematics is based.

Mathographical Remarks

In Chapter 12, we shall define a set $V_{\omega+\omega}$ such that the standard model $\langle V_{\omega+\omega}, \overline{\in}\rangle$ satisfies all axioms of ZFC except (A7). Moreover, it will be shown that for all $n \in \omega$, the n-th successor $S^{(n)}(\omega) \in V_{\omega+\omega}$, but $S = \{S^{(n)}(\omega) : n \in \omega\} \notin V_{\omega+\omega}$. It follows that if you managed to solve Exercise 24, then your solution must contain an error.

Set theory without the Axiom of Choice has been intensely studied by mathematicians who were suspicious about the nonconstructive character of (A9). The axiom system ZF, which is obtained by deleting the Axiom of Choice from the list of axioms presented in this chapter, provides a possible framework for mathematics without the Axiom of Choice. In Section 9.2 we give an overview of mathematics without the Axiom of Choice.

It is possible to develop set theory without the Foundation Axiom. An interesting approach was suggested by P. Aczel in *Non-Well-Founded Sets,* CLSI Lecture Notes, Vol. 14, CLSI Publications, Stanford, California, 1988. Using the so-called Anti-Foundation Axiom instead of the Foundation Axiom, he developed an alternative approach to set theory which is of significant interest to computer scientists.

CHAPTER 8

Classes

Assume that $\langle M, E\rangle \models$ ZFC, and let $\varphi(x)$ be a formula with one free variable. Let $\mathbf{X} = \varphi^{\langle M,E\rangle}$, where $\varphi^{\langle M,E\rangle} = \{x \in M : \langle M, E\rangle \models \varphi(x)\}$. Then $\mathbf{X}$ is a collection of elements of M, but $\mathbf{X}$ does not necessarily correspond to an element x of M. In the language of the model $\langle M, E\rangle$, we can refer to $\mathbf{X}$ by writing "$\varphi(x)$." Nevertheless, mathematicians find it convenient to use expressions like "$x \in \mathbf{X}$" (which, strictly speaking, belong to the metalanguage) as abbreviations for "$\varphi(x)$." If there is no x in M such that $\mathbf{X}$ corresponds to x, then $\mathbf{X}$ is called a *proper class.* If we are not sure whether or not $\mathbf{X}$ corresponds to some $x \in M$, then we call $\mathbf{X}$ simply a *class.* It follows that if $\mathbf{X}$ is a proper class, then expressions like "$\mathbf{X} \in y$" or "$\mathbf{X} \in \mathbf{Y}$" do not abbreviate any formulas of the language of the model, and should never be used.

In principle, this is all there is to proper classes. Nevertheless, many mathematicians find the notion of a proper class confusing. How come? In its capacity as a foundation of mathematics, ZFC is not usually looked upon as describing properties of a somewhat arbitrary model, but as a tool for discovering the truth about THE UNIVERSE OF SETS. The underlying view is that THE UNIVERSE OF SETS exists independently of human intelligence, in much the same way as the physical universe exists independently of us. This philosophical standpoint is called *Platonism.* According to the Platonist view, mathematicians *discover* properties of THE UNIVERSE OF SETS. According to opposing views, mathematicians just *invent* models of ZFC.

We do not wish to discuss the relative merits of Platonism and its alternatives. There is no way of deciding whether or not Platonism is the "correct" philosophy. That's how it is with philosophical views in general. However, we cannot ignore the observation that *when working on a mathematical problem,* all mathematicians are Platonists. A mathematician who is studying properties of the real line will think of this process as *discovering* truths about an object that exists independently of the human mind. When not trying to prove a theorem, this same mathematician may view the reals as a pure abstraction invented for the sole purpose of describing certain properties of measurements. While the latter view may be this mathematician's personal philosophy of mathematics, it does not reflect his personal experience as a theorem-prover.[1]

When treating sets as elements of THE UNIVERSE OF SETS, a mathematician is still able to picture the collection of all sets that share a certain property. But he knows that collections like the collection of all sets x such that $x \notin x$ cannot be members of THE UNIVERSE OF SETS. Thus, he needs a separate name for collections of sets which are not themselves sets. Hence the notion of a proper class.

[1]We are describing the way a majority of us think. There may be rare exceptions.

As we mentioned above, when working with classes, one must exercise a lot of intellectual self-restraint. For instance, one must never try to collect two proper classes into a set. At first glance, this seems to be an absurd restriction. Why should forming a pair $\{\mathbf{X}, \mathbf{Y}\}$ be any different from building $\{\emptyset, \{\emptyset\}\}$? Is it possible to maintain that a pair $\{\mathbf{X}, \mathbf{Y}\}$ does not belong to THE UNIVERSE OF *ALL* SETS?

Yes. Recall what we said in the Introduction about the architect's view of set theory. Sets are built from preexisting building blocks, at moments of an abstract "set-theoretic timeline." This "timeline" will be discussed in Chapter 12, where we shall also prove that a class $\mathbf{X}$ is proper if and only if at any moment in set-theoretic time, there is some $x \in \mathbf{X}$ that has not yet been constructed. Therefore, in particular, if at least one of $\mathbf{X}, \mathbf{Y}$ is a proper class, then the pair $\{\mathbf{X}, \mathbf{Y}\}$ could not have been formed at any moment in set-theoretic history.

If one does not entertain the notion of THE UNIVERSE OF SETS in all its metaphysical seriousness, then one can imagine that mathematics takes place inside a model $\langle M_0, \in_0 \rangle$ of set theory. Proper classes of this model may be understood as sets of some larger model $\langle M_1, \in_1 \rangle$, and so on. Of course, when this approach is taken, one must sooner or later face the problem of how to handle the collection of all these nested models. Again, we see how human creativity in mathematics can transcend any formal system. A different aspect of the same phenomenon has already been discussed in Section 6.2. There, we saw that mathematics can be based on a fixed set of axioms only if one is willing to restrict the use of certain obvious truths in mathematical arguments. Here, we see that the adoption of a fixed axiom system also forces us to refrain from the use of certain rather plausible constructions.

While mathematicians generally appreciate the sense of certainty that comes from an axiomatic foundation, most of us are usually unaware of the restrictions imposed by an axiomatic framework. This is so because ZFC is broad enough to accomodate almost any product of a mathematician's imagination. The notion of a proper class straddles the boundary between concepts that can be formalized within ZFC, and notions that transcend this system. Thus, when working with proper classes, one comes face to face with the restrictions imposed by the axiomatic framework, and this may be a rather unsettling experience.

Now let us consider some important proper classes. We write "$x \in \mathbf{V}$" as an abbreviation for "$x = x$." Thus, $\mathbf{V}$ is the class of all sets. According to the Platonist view, $\mathbf{V}$ = THE UNIVERSE OF SETS. $\mathbf{V}$ is a proper class: If $\mathbf{V}$ would correspond to a set v, then we could use an instance of the Separation Axiom Schema to form the set $y = \{x \in v : x \notin x\}$. As you already know from the Introduction, the assumption that y is a set leads to Russell's Paradox. Therefore, the assumption that $\mathbf{V}$ corresponds to a set v yields a contradiction; and we conclude that $\mathbf{V}$ is a proper class.

If $\mathbf{X}$ is the class of all x such that $\varphi(x)$, then the formula $\varphi(x)$ is called a *representation of the class* $\mathbf{X}$. Representations of classes are not uniquely determined. For example, the formula "$x = x$" is a representation of the class $\mathbf{V}$, but it is not the only representation of this class.

EXERCISE 1(G): Find two other representations of the class $\mathbf{V}$.

A class $\mathbf{X}$ that consists entirely of ordered pairs $\langle x, y\rangle$ is called a *relational class.*[2] Relational classes act like binary relations on $\mathbf{V}$. Therefore, one may define relational classes as follows:

Definition 1: A class $\mathbf{X}$ is a *relational class* if $\mathbf{X} \subseteq \mathbf{V} \times \mathbf{V}$.

Are we allowed to write like this? When using proper classes we must make sure that all expressions written in this terminology correspond to expressions of L_S.

EXERCISE 2(G): Suppose $\varphi(x)$ is a representation of a class $\mathbf{X}$. Convince yourself that the expression "$\mathbf{X} \subseteq \mathbf{V} \times \mathbf{V}$" of the metalanguage corresponds to the following expression of L_S: "$\forall x\,(\varphi(x) \rightarrow \exists y, z\,(x = \langle y, z\rangle))$."

EXERCISE 3(G): Suppose $\varphi(x)$ is a representation of a class $\mathbf{X}$, and $\psi(x)$ is a representation of a class $\mathbf{Y}$.

(a) Show that $\mathbf{X} \cap \mathbf{Y}$, $\mathbf{X} \cup \mathbf{Y}$, $\mathbf{X} \backslash \mathbf{Y}$, $\mathbf{X} \times \mathbf{Y}$ and $\bigcup \mathbf{X}$ are also classes. *Hint:* Find suitable representations.

(b) Find an expression of L_S that corresponds to "$\mathbf{X} \subseteq \mathbf{Y}$."

If $\mathbf{X}$ and $\mathbf{Y}$ are classes such that $\mathbf{X} \subseteq \mathbf{Y}$, then we call $\mathbf{X}$ *a subclass of* $\mathbf{Y}$.

We have already seen two examples of proper relational classes:
The class $\mathbf{EQ} = \{\langle x, y\rangle : x \approx y\}$, and the class $\mathbf{OI} = \{\langle x, y\rangle : x, y$ are p.o.'s and $x \cong y\}$. Note that if $\langle x, y\rangle \in \mathbf{OI}$, then x and y are themselves ordered pairs.

EXERCISE 4(G): (a) Show that $\bigcup\bigcup \mathbf{EQ} = \mathbf{V}$.

(b) Use Exercise 3(a) and the Union Axiom to conclude that the class $\mathbf{EQ}$ is proper.

EXERCISE 5(PG): Show that the class $\mathbf{OI}$ is proper.

Representations of proper classes may contain parameters. For example, for each $x \in \mathbf{V}$ let $\mathbf{CARD}(x) = \{y : x \approx y\}$.

EXERCISE 6(PG): Show that if $x \neq \emptyset$, then $\bigcup \mathbf{CARD}(x) = \mathbf{V}$.

It follows that unless x is the empty set, $\mathbf{CARD}(x)$ is a proper class. In Chapter 3 we encouraged you to interpret the symbol $|x|$ as $\mathbf{CARD}(x)$.[3] This approach works well up to a point; but it has its limitations. Here is one: In Chapter 3 we also defined the operations of addition, multiplication, and exponentiation of cardinals. To be specific, let us now consider only addition of cardinals. What kind of an object is this operation? It should be a function of two arguments. But domains of functions are sets; and intuition tells us that the domain of this operation would most likely turn out to be a proper class. Okay, let us try to formalize addition of cardinals as a *functional class* $\mathbf{PLUS}$. Functional classes will be discussed in a moment; for now it suffices to know that functional classes are relational classes with special properties. If $\mathbf{PLUS}$ is going to be a relational class, then its elements

[2] We really should call such classes "binary relational classes," but since all relational classes considered in this book are binary, there is no need for the extra adjective.

[3] Of course, we used a different terminology.

will be ordered pairs $\langle x, y\rangle$, where y is going to be a cardinal, and x is going to be an ordered pair of cardinals. But, if cardinals are proper classes, then we are not even allowed to form unordered pairs of cardinals, let alone ordered pairs and pairs of ordered pairs!

A similar remark applies to order types: If order types are proper classes, then the operations of addition and multiplication of order types cannot be defined in the present framework.

Too bad. Before we can show how to overcome this obstacle, we need to take a closer look at functional classes.

Definition 2: A relational class $\mathbf{F}$ is a *functional class* if

$$\forall x, y, z\,(\langle x, y\rangle \in \mathbf{F} \wedge \langle x, z\rangle \in \mathbf{F} \rightarrow y = z).$$

A functional class $\mathbf{F}$ is *injective* if

$$\forall x, y, z\,(\langle x, z\rangle \in \mathbf{F} \wedge \langle y, z\rangle \in \mathbf{F} \rightarrow x = y).$$

For example, the collection $\mathbf{S} = \{\langle x, x \cup \{x\}\rangle : x \in \mathbf{V}\}$ is a functional class. It acts like a function that assigns to each set x its successor $x \cup \{x\}$; but the collection is too big to form a set.

EXERCISE 7(G): Show that the class $\mathbf{S}$ is proper.

EXERCISE 8(G): Let $\psi(w)$ be a formula of L_S with one free variable. Show that there is a formula $\varphi(x, y)$ with two free variables such that

$$\text{ZFC} \vdash \varphi(x, y) \leftrightarrow \psi(w) \wedge w = \langle x, y\rangle.$$

Let $\psi(w)$ be a representation of a relational class $\mathbf{F}$, and let $\varphi(x, y)$ be the formula of Exercise 8. Then the class $\mathbf{F}$ is functional if and only if

$$\forall x, y, z\,(\varphi(x, y) \wedge \varphi(x, z) \rightarrow y = z).$$

EXERCISE 9(G): (a) Find a suitable definition of the *domain* of a functional class.

(b) Prove that if $\mathbf{F}$ is a functional class, then $dom(\mathbf{F})$, the domain of $\mathbf{F}$, is also a class. *Hint:* Find a representation of $dom(\mathbf{F})$.

If $\mathbf{F}$ is a functional class, and a is a subset of the domain of $\mathbf{F}$, denote $\mathbf{F}[a] = \{y : \exists x \in a\,(\langle x, y\rangle \in \mathbf{F})\}$. This terminology allows us to reformulate the Replacement Axiom Schema as follows:

(A7′) If $\mathbf{F}$ is a functional class, and a is a subset of the domain of $\mathbf{F}$, then $\mathbf{F}[a]$ is also a set.

EXERCISE 10(G): Convince yourself that (A7′) really does say the same thing as (A7).

EXERCISE 11(PG): Why does the new formulation (A7′) of the Replacement Axiom Schema not allow us to condense this axiom schema into a single axiom?

The terminology of functional classes makes it easier to see why the Replacement Axiom Schema is not a consequence of the other axioms of ZFC. Consider the following statement:

(A7$''$) If $\mathbf{F}$ is a functional class, and a is a subset of the domain of $\mathbf{F}$, then $\mathbf{F}|a = \{\langle x, y\rangle \in \mathbf{F} : x \in a\}$ is a function.

EXERCISE 12(PG): (a) Show that (A7$''$) corresponds to a collection of theorems of ZFC.

(b) Show that all instances of the Replacement Axiom Schema are provable in ZFC $\backslash\{$(A7)$\}$ from (A7$''$).

EXERCISE 13(PG): Using the terminology of functional classes and (A7$''$), write your own version of the paragraph which immediatly follows Exercise 15 of the previous chapter.

Now let us return to the question of how one could define the operation of cardinal addition without transcending the framework of sets and classes. As we already indicated in Chapter 3, in axiomatic set theory cardinals are not defined as proper equivalence classes, but rather as chosen representatives of these classes. In Chapter 11 we shall show you how one can define a functional class **CAR** such that for all $x, y, z \in \mathbf{V}$:

(1) If $\langle x, y\rangle \in \mathbf{CAR}$, then $x \approx y$; and

(2) If $\langle x, z\rangle \in \mathbf{CAR}$ and $x \approx y$, then $\langle y, z\rangle \in \mathbf{CAR}$.

For each x, the symbol $|x|$ will denote the unique y such that $\langle x, y\rangle \in \mathbf{CAR}$.

EXERCISE 14(G): Reread Footnote 3 of Chapter 3 and the sentence it refers to. Convince yourself that the unique y with $\langle x, y\rangle \in \mathbf{CAR}$ acts like the "cardinal numbers" we described in Chapter 3.

We conclude this chapter with an intriguing question:

Question 3: Is there a relational class $\mathbf{W}$ that wellorders the whole universe $\mathbf{V}$?

Let us first convince ourselves that this question is meaningful.

Definition 4: A relational class $\mathbf{W}$ is said to *wellorder a class* $\mathbf{X}$ if it satisfies the following conditions:

(W1) $\forall x \in \mathbf{X}\,(\langle x, x\rangle \in \mathbf{W})$;

(W2) $\forall x, y \in \mathbf{X}\,(\langle x, y\rangle \in \mathbf{W} \wedge \langle y, x\rangle \in \mathbf{W} \rightarrow x = y)$;

(W3) $\forall x, y, z \in \mathbf{X}\,(\langle x, y\rangle \in \mathbf{W} \wedge \langle y, z\rangle \in \mathbf{W} \rightarrow \langle x, z\rangle \in \mathbf{W})$;

(W4) $\forall x, y \in \mathbf{X}\,(\langle x, y\rangle \in \mathbf{W} \vee \langle y, x\rangle \in \mathbf{W})$;

(W5) $\forall x \in \mathbf{X}\,(x \cap \mathbf{X} \neq \emptyset \rightarrow \exists y \in x \cap \mathbf{X}\, \forall z \in x \cap \mathbf{X}\,(\langle y, z\rangle \in \mathbf{W}))$.

A relational class $\mathbf{W}$ is said to be *set-like* if

(W6) $\forall x \exists y\,(y = \{z : \langle z, x\rangle \in \mathbf{W}\})$.

Condition (W6) says that the collection of $\mathbf{W}$-predecessors of any set x again forms a set.

If $\mathbf{W}$ is a relational class that wellorders a class $\mathbf{X}$, then there must exist a formula $\varphi(x, y)$ such that $\forall x, y\,(\langle x, y\rangle \in \mathbf{W} \leftrightarrow \varphi(x, y))$.

EXERCISE 15(G): Assume that $\varphi(x, y)$ is such a formula, and rewrite conditions (W1)–(W6) in terms of $\varphi(x, y)$ instead of $\mathbf{W}$.

A relational class that wellorders the universe $\mathbf{V}$ is often called a *definable wellorder of the universe.* In Chapter 12 we shall see that the existence of a definable, set-like wellorder of the universe is relatively consistent with ZFC. Classes that satisfy (W1)–(W6) allow one to construct all sorts of other useful objects.

EXERCISE 16(PG): Let $\mathbf{PO}$ be the class of all p.o.'s. Assume that there exists a set-like relational class $\mathbf{W}$ that wellorders the universe. Prove under this assumption that there exists a functional class $\mathbf{OT}$ that satisfies:
(i) If $\langle x, y\rangle \in \mathbf{OT}$, then x and y are p.o.'s such that $x \cong y$;
(ii) The domain of $\mathbf{OT}$ is the class $\mathbf{PO}$;
(iii) $\forall x, y \in \mathbf{PO}\,(x \cong y \wedge \langle x, z\rangle \in \mathbf{OT} \rightarrow \langle y, z\rangle \in \mathbf{OT})$.

Let $\mathbf{OT}$ be a functional class that satisfies (i)–(iii). We can use this class to define for an arbitrary p.o. $\langle X, \preceq\rangle$ the order type $ot(\langle X, \preceq\rangle)$ as the unique p.o. $\langle Y, \leq\rangle$ such that $\langle\langle X, \preceq\rangle, \langle Y, \leq\rangle\rangle \in \mathbf{OT}$. This approach is exactly the same as the one for defining $|x|$ which we outlined earlier in this chapter. Unfortunately, a negative answer to Question 3 is also relatively consistent with ZFC. We shall see in Chapter 10 how one can give in ZFC a rigorous definition of order types of strict w.o.'s in such a way that $ot(\langle X, \prec\rangle)$ will be a strict w.o. $\langle Y, <\rangle$ such that $\langle Y, <\rangle \cong \langle X, \prec\rangle$. In Chapter 12, we shall show how one can define order types of arbitrary p.o.'s in ZF alone in such a way that the order types will be sets rather than proper classes.

A relational class $\mathbf{W}$ is said to be *wellfounded* if

$$\forall x\,(x \neq \emptyset \rightarrow \exists y \in x \forall z\,(y \neq z \rightarrow \langle z, y\rangle \notin \mathbf{W})).$$

The following theorem generalizes Theorem 4.25:

Theorem 5: *Let $\mathbf{Z} \neq \emptyset$ and let $\mathbf{W} \subseteq \mathbf{Z} \times \mathbf{Z}$ be a wellfounded set-like class. Let $\mathbf{G}$ be a functional class such that $dom(\mathbf{G})$ consists of all pairs of the form $\langle f, z\rangle$, where $z \in \mathbf{Z}$ and f is a function with domain $I_{\mathbf{W}}(z)$. Then there exists exactly one functional class $\mathbf{F} : \mathbf{Z} \rightarrow \mathbf{V}$ such that*

$$\text{(R)} \qquad \mathbf{F}(z) = \mathbf{G}(\mathbf{F}|I_{\mathbf{W}}(z), z) \quad \textit{for all } z \in \mathbf{Z}.$$

EXERCISE 17(PG): Prove Theorem 5. *Hint:* Consider representations of $\mathbf{Z}$, $\mathbf{W}$, $\mathbf{G}$ and apply Theorem 7.3.

Theorem 4.21 can also be generalized to classes.

EXERCISE 18(PG): Formulate the analogue of Theorem 4.21 for classes.

Mathographical Remark

The axiom systems NBG (von Neumann-Bernays-Gödel set theory) and MK (Morse-Kelley set theory) are alternatives to ZFC. They are formulated in a formal language that contains special symbols for proper classes. In particular, (A7′) and (A7″) can be formulated as single axioms in this language. A thorough discussion of alternative systems on which set theory can be founded is contained in

A. Fraenkel, Y. Bar-Hillel, and A. Lévy, *Foundations of Set Theory*, North-Holland, Amsterdam, 1973.

CHAPTER 9

Versions of the Axiom of Choice

The proofs of several theorems stated in this chapter will be given only in Chapters 10 and 11. If you do not like such reversals, you may want to adopt the following order of reading: Chapter 10, Section 9.1, Chapter 11, and then Section 9.2. On the other hand, the present chapter provides the motivation for developing a theory of ordinals in ZF rather than ZFC. If you prefer to be motivated before doing a piece of work, we recommend that you read the chapters in the order of our presentation.

The last two sections are included to satisfy the reader's curiosity; they do not contain material that will be used later in this text. Section 9.3 is intended to give a glimpse at a rapidly developing topic of modern set theory. This unit is called a "Section," and it contains exercises, definitions and theorems; but its character is more that of an oversized mathographical remark. Section 9.4 is entirely devoted to the proof of one of the theorems discussed in Section 9.2.

9.1. Statements Equivalent to the Axiom of Choice

EXERCISE 1(PG): Is the Fundamental Theorem of Calculus equivalent to the Fundamental Theorem of Algebra?

In mathematical texts, one often finds assertions like "Theorem A is equivalent to Theorem B." Much of the time, this is supposed to mean "Theorem B trivially follows from Theorem A, and Theorem A is an easy consequence of Theorem B." As we have seen in Chapter 6, the notion of an "easy consequence" is not at all easy to formalize. We shall never use the word "equivalent" as a respectably sounding synonym for "trivially follow from each other." Instead, we call two sentences φ and ψ equivalent if it can be proved that either both φ and ψ are true, or both φ and ψ are false.

EXERCISE 2(G): Is the latter definition of "equivalent" rigorous?

Let us reconsider Exercise 1 in light of our concept of equivalence. Since both the Fundamental Theorem of Calculus and the Fundamental Theorem of Algebra are *theorems,* they are both true statements, and therefore they are equivalent.

But wait a minute. These are theorems of different branches of mathematics. If we had two languages L_A (of algebra) and L_C (of calculus), a first-order theory T_A in L_A and a first-order theory T_C in L_C such that $T_A \vdash \varphi_A$ and $T_C \vdash \varphi_C$ (where φ_A, φ_C stand for the Fundamental Theorems of Algebra and Calculus), would the definition of "equivalence" given above Exercise 2 imply that φ_A is equivalent to φ_C?

EXERCISE 3(G): Show that if φ_0, φ_1 are arbitrary sentences of any two languages L_0, L_1, then there are theories T_0, T_1 in these languages such that $T_0 \vdash \varphi_0$ and $T_1 \vdash \varphi_1$.

Now our definition of equivalence starts to look useless. We must be more specific.

Definition 1: Let T be a theory in a formal language L, and let φ, ψ be sentences of L. We say that *φ and ψ are equivalent in T* if

$$T \vdash \varphi \leftrightarrow \psi.$$

Equipped with this definition, let us return once more to Exercise 1. Both the Fundamental Theorem of Algebra φ_A and the Fundamental Theorem of Calculus φ_C can be expressed as sentences of L_S, at least in principle. (Or did you really complete Exercise 1 of the Introduction?) Moreover, ZFC $\vdash \varphi_A$ and ZFC $\vdash \varphi_C$. Therefore, ZFC $\vdash \varphi_A \leftrightarrow \varphi_C$; and it follows that the Fundamental Theorem of Calculus is equivalent in ZFC to the Fundamental Theorem of Algebra. For very much the same reason, the Prime Number Theorem is equivalent in ZFC to the sentence "$2 + 2 = 4$," and the Axiom of Choice is equivalent in ZFC to the Empty Set Axiom.

These rather uninspiring examples notwithstanding, the notion of equivalence in ZFC does have its uses. For example, if φ is a sentence whose relative independence of ZFC has been established, then proving that a sentence ψ is equivalent in ZFC to φ is a way of establishing that ψ is also relatively independent of ZFC. The latter is often much easier than finding a separate independence proof for ψ.

Moreover, it frequently happens that mathematicians are unable to prove or disprove certain conjectures φ and ψ, but they can show that the two conjectures are equivalent in ZFC. This kind of result can be valuable, especially if φ and ψ belong to different subdisciplines of mathematics, or if there is strong empirical evidence for the truth or falsity of one of these sentences. For example, at the time of this writing, one of the most important unsolved mathematical problems is the famous P-NP-problem. It concerns two sets of computable functions which are commonly denoted by P and NP. Most contemporary mathematicians believe that P $\neq$ NP, but so far nobody has been able to prove this inequality. However, there are literally hundreds of mathematically meaningful sentences which are known to be equivalent in ZFC to the inequality P $\neq$ NP. Most of these sentences assert that certain problems are too complex to be solvable by a computer in any reasonable amount of computing time, but some of them are statements which one would not expect to have any connections to computational complexity.

In the remainder of this chapter, whenever we write "equivalent," we mean *equivalent in ZF*. As the title suggests, we shall present several sentences that are equivalent to the Axiom of Choice. We shall frequently use the following properties of the provability relation $\vdash$:

Fact 2: *Let T be any first-order theory in a language L, and let φ and ψ be sentences in L. Then*

(1) $T \vdash \varphi \leftrightarrow \psi$ *iff* $T \vdash \varphi \rightarrow \psi$ *and* $T \vdash \psi \rightarrow \varphi$;

(2) $T \cup \{\varphi\} \vdash \psi$ *iff* $T \vdash \varphi \rightarrow \psi$.

EXERCISE 4(PG): Let φ be a sentence of L_S, and let ZF + φ be the theory one obtains by adding φ to the axioms of ZF (i.e., replacing axiom (A9) of ZFC by φ). Convince yourself that φ is equivalent in ZF to the Axiom of Choice if and only if $(\mathrm{ZF} + \varphi)^{\vdash} = \mathrm{ZFC}^{\vdash}$.

Let us begin with a slight modification of (A9).

(AC) Let $\{A_i : i \in I\}$ be an indexed family of nonempty sets. Then $\prod_{i \in I} A_i \neq \emptyset$.

EXERCISE 5(G): Show that (A9) and (AC) are equivalent in ZF.

Both (A9) and (AC) are commonly referred to as the "Axiom of Choice," and we shall follow this practice. After all, the difference between the two statements is one of style rather than content. If $\mathcal{A} = \{A_i : i \in I\}$ is an indexed family and $f \in \prod_{i \in I} A_i$, then f will be called a *choice function for* $\mathcal{A}$. If A is a family of pairwise disjoint sets and $b \subseteq \bigcup A$ is such that $|b \cap a| = 1$ for each $a \in A$, then b will be called a *selector for* A.[1]

Consider Theorem 7.5, the Principle of Recursive Constructions over Well-founded Sets, which henceforth will be abbreviated (RCW). Since Theorem 7.5 was proved in ZFC, we know that ZFC $\vdash$ (RCW). It follows from Fact 2 that ZF $\vdash$ (A9) $\to$ (RCW). Now we prove the converse implication. In order to indicate that Theorem n is proved in ZF rather than in ZFC, we shall write "Theorem n (ZF): ... ."

Theorem 3 (ZF): (RCW) $\to$ (AC).

Proof: Assume that (RCW) holds, and let $\mathcal{A} = \{A_z : z \in Z\}$ be an indexed family of nonempty sets. If $Z = \emptyset$, then $\emptyset$ is a choice function for $\mathcal{A}$; so we may without loss of generality assume that $Z \neq \emptyset$. We want to construct a function F with domain Z such that $F(z) \in A_z$ for all $z \in Z$. In order to use (RCW), we need a set X, a strictly wellfounded relation W on Z, and a function G^* as in the assumptions of (RCW). $\bigcup \mathcal{A}$ is a natural candidate for X, and the simplest possible strictly wellfounded relation on Z, the empty set, looks like a good candidate for W.

EXERCISE 6(G): Convince yourself that if $W = \emptyset$, then W is a strictly well-founded relation on Z, and every $z \in Z$ is W-minimal.

Hence our choice of W allows us to apply (RCW) to the function G^* that is defined by: $G^*(\emptyset, z) = A_z$ for $z \in Z$.

EXERCISE 7(G): Let F be a function that satisfies conditions (i) and (ii) of (RCW) for these choices of Z, X, W, and G^*. Verify that $F \in \prod_{z \in Z} A_z$. $\square$

---

[1]Many authors use the terms "selector" and "choice function" interchangably. Some fields of mathematics have their own special terminology: In combinatorics, selectors are called "transversals," and descriptive set theorists speak of "uniformizations."

Theorems 3 and 7.5 together imply that (RCW) and (AC) are equivalent in ZF.

Given an indexed family $\mathcal{A} = \{A_i : i \in I\}$ of nonempty sets, it is not always necessary to invoke the Axiom of Choice in order to prove that $\prod_{i \in I} A_i \neq \emptyset$. For example, if I is finite, then it is possible to construct an $f \in \prod_{i \in I} A_i$ using only the axioms of ZF.

EXERCISE 8(PG): List the axioms that are needed to find such an f for a given product $\prod_{i \in I} A_i$ over a finite set of indices I.
Hint: Distinguish two cases: when I is empty, and when I is nonempty.

Also, if each of the sets A_i has exactly one element, then there isn't anything to choose from, and we do not need the Axiom of Choice to find a function $f \in \prod_{i \in I} A_i$. Similarly, if each of the A_i's contains an element that is somehow special in a uniformly describable way, then one can show the existence of an $f \in \prod_{i \in I} A_i$ from the axioms of ZF alone.

EXERCISE 9(PG): "Being special in a uniformly describable way" means that there are a formula $\varphi(v_0, v_1, \dots, v_k)$ of L_S with $k+1$ free variables for some $k \in \omega$ and parameters $b_1, \dots, b_{k-1}$ such that $\forall i \in I \exists! a \in A_i\, \varphi(a, b_1, \dots, b_{k-1}, i)$. Suppose that such a formula φ exists, and show from the axioms of ZF that there exists a function $f \in \prod_{i \in I} A_i$.

The situation described in Exercise 9 occurs for example if $\bigcap_{i \in I} A_i \neq \emptyset$. Then we can pick some $b \in \bigcap_{i \in I} A_i$ and let $\varphi(a, b)$ be the formula "$a = b$." A more mundane example is the situation where each A_i is a pair of shoes. Then we can choose $\varphi(a)$: "a is a left shoe." Note that we do not have recourse to a similar device if each A_i is a pair of socks. One can show that ZF does not prove the existence of a choice function $\prod_{i \in I} A_i$ even in all cases where each A_i is a two-element set.

The most important instance of the situation described in Exercise 9 occurs if there exists a wellorder relation on $\bigcup \mathcal{A}$.

Corollary 4 (ZF): *Suppose $\mathcal{A} = \{A_i : i \in I\}$ is an indexed family of nonempty sets. If there exists a wellorder relation on $\bigcup \mathcal{A}$, then $\prod_{i \in I} A_i \neq \emptyset$.*

Proof: Let $\mathcal{A}$ be as in the assumptions, and let $\preceq$ be a wellorder relation on $\bigcup \mathcal{A}$. Let $\varphi(a, \preceq, i)$ be the formula: "a is the $\preceq$-minimum element of A_i." By Exercise 9, there exists an $f \in \prod_{i \in I} A_i$. □

Note that it does not suffice to require that each of the sets A_i individually carries a wellorder relation: If each A_i has two elements, then each A_i separately can be wellordered. But as we mentioned above, this does not imply the existence of a choice function for $\{A_i : i \in I\}$.

Now consider the following statement.

(WO) For each set X there exists an $R \subseteq X^2$ such that $\langle X, R \rangle$ is a w.o.

The statement (WO) is called the *Wellorder Principle* or *Zermelo's Theorem.* The following is an immediate consequence of Corollary 4:

Corollary 5: ZF $\vdash$ (WO) $\rightarrow$ (AC).

We encountered (WO) already in Part I: In Exercise 2.13 it was shown that the following is a consequence of (WO):

(DIS) For every indexed family $\{A_i : i \in I\}$ there exists a family $\{B_i : i \in I\}$ of pairwise disjoint sets such that $B_i \subseteq A_i$ for all $i \in I$ and $\bigcup_{i \in I} B_i = \bigcup_{i \in I} A_i$.

EXERCISE 10(R): Show that ZF $\vdash$ (DIS) $\rightarrow$ (A9).

We are going to show that ZF $\vdash$ (AC) $\rightarrow$ (WO), and hence, that the statements (AC), (WO) and (DIS) are all equivalent in ZF. Along the way, we shall derive some other useful versions of the Axiom of Choice.

Recall that a *chain* in a p.o. $\langle X, \preceq \rangle$ is a subset $C \subseteq X$ that is linearly ordered by $\preceq$. A chain C is *maximal* if no proper superset $D \supset C$ is a chain in $\langle X, \preceq \rangle$. *Hausdorff's Maximal Principle* is the following statement:

(HMP) Every chain in a p.o. can be extended to a maximal chain.

We split the proof of (HMP) into several stages. First we formulate Theorem 6 which is related to (HMP) in a manner similar to the way Theorem 4.25 is related to Theorem 4.26. Next we show how Theorem 6 implies (HMP). This part of the argument uses the Axiom of Choice. Theorem 6 itself is a theorem of ZF. Its proof will be given in the next chapter.

Let $\mathcal{C}$ be a family of sets.[2] We say that $\mathcal{C}$ *is closed under unions of subfamilies that are linearly ordered by inclusion* if $\forall \mathcal{B} \subseteq \mathcal{C}\, (\forall A, B \in \mathcal{B}\, (A \subseteq B \vee B \subseteq A) \rightarrow \bigcup \mathcal{B} \in \mathcal{C})$.

Theorem 6 (ZF): *Let X be a set, let $\mathcal{C}$ be a family of subsets of X that is closed under unions of subfamilies that are linearly ordered by inclusion, and let $f : \mathcal{C} \rightarrow \mathcal{C}$ be a function such that $f(C) \supseteq C$ for all $C \in \mathcal{C}$. Then there exists a $C \in \mathcal{C}$ such that $f(C) = C$.*

Proof of (HMP) from Theorem 6: Let $\langle X, \preceq \rangle$ be a p.o., and let $C_0 \subseteq X$ be a chain in this p.o. Denote:

$$\mathcal{C} = \{C : C_0 \subseteq C \subseteq X \wedge C \text{ is a chain in } \langle X, \preceq \rangle\}.$$

EXERCISE 11(G): Show that $\mathcal{C}$ is closed under unions of subfamilies that are linearly ordered by inclusion.

For each $C \in \mathcal{C}$, define a set $\mathcal{A}_C$ as follows: If C is a maximal chain in $\langle X, \preceq \rangle$, then $\mathcal{A}_C = \{C\}$. If C is not maximal, then $\mathcal{A}_C$ is the set of chains in $\langle X, \preceq \rangle$ that

[2] Since we are now practicing axiomatic set theory, there is no real need to distinguish between sets and families of sets: Every set is a set of sets, i.e., a family. But there is no harm in using the word "family" as a synonym for "set;" and sometimes an emphasis of the fact that the elements of our set are themselves sets may aid intuition.

properly extend[3] C. By the Axiom of Choice, there exists a function $f \in \prod_{C \in \mathcal{C}} \mathcal{A}_C$. Fix such a function f and note that f satisfies the assumptions of Theorem 6. Thus, there exists a $C \in \mathcal{C}$ such that $f(C) = C$. Then C is a maximal chain in $\langle X, \preceq \rangle$ that extends C_0: If C were not maximal, then $f(C)$ would be a proper extension of C, which it is not. Since C_0 was arbitrary, we have proved (HMP). □

An *upper bound* for a chain C in a p.o. $\langle X, \preceq \rangle$ is an element $x \in X$ such that $c \preceq x$ holds for every $c \in C$. Note that an upper bound for a chain C may or may not be an element of C. The following statement, known as *Zorn's Lemma,* is a close relative of Hausdorff's Maximal Principle:

(ZL) Let $\langle X, \preceq \rangle$ be a p.o. such that each chain in X has an upper bound. Then $\langle X, \preceq \rangle$ has a maximal element.

Zorn's Lemma was first formulated and proved by the Polish mathematician Kazimierz Kuratowski. Zorn discovered it later independently.

Theorem 7 (ZF): (HMP) $\rightarrow$ (ZL).

Proof: Assume that (HMP) holds, and let $\langle X, \preceq \rangle$ be a p.o. such that every chain in X has an upper bound. Note that $\emptyset$ is a chain in X. By (HMP), there exists a maximal chain C in X that extends $\emptyset$. Fix such a maximal chain C, and let x be an upper bound of C. Assume that x is not maximal in $\langle X, \preceq \rangle$, and let $y \in X$ be such that $x \prec y$. Then $c \prec y$ for all $c \in C$, and thus the set $C \cup \{y\}$ is a chain in X, which contradicts maximality of C. Hence, x must be maximal in $\langle X, \preceq \rangle$, and we have proved Theorem 7. □

Among all statements that are equivalent in ZF to (AC), the two most important ones are Zorn's Lemma and the Wellordering Principle. Since applications of (ZL) and (WO) are found in most branches of mathematics, every mathematician needs to know how to make good use of these theorems.

Most proofs using Zorn's Lemma follow a fixed pattern. We are going to illustrate this pattern with a few important examples.

Our first example involves the notion of an *ultrafilter*. Ultrafilters play a pivotal role in topology and mathematical logic, and will also be extensively used in Volume II.

Definition 8: Let $x \neq \emptyset$. A *filter* F *on* x is a nonempty collection of subsets of x such that

(i) F is *closed under supersets,* that is, $\forall y \in F\, \forall z \subseteq x\, (y \subseteq z \rightarrow z \in F)$.

(ii) F is *closed under finite intersections,* i.e., $\forall H \subseteq F\, (H \text{ is finite} \rightarrow \bigcap H \in F)$.

A filter F is *proper* if $\emptyset \notin F$.

EXERCISE 12(G): Let x be a nonempty set.

(a) Show that if F is a filter on x, then $x \in F$.

(b) Show that $\mathcal{P}(x)$ is the only improper filter on x.

(c) Show that if we replace (ii) of Definition 7 by the condition:

(ii)′ $\forall y, z \in F\, (y \cap z \in F)$,

then we get an equivalent definition.

[3] We say that a chain D *properly extends* a chain C if C is a proper subset of D.

(d) Show that for every subset y of x the family $\check{y} = \{z \subseteq x : y \subseteq z\}$ is a filter on x.

(e) Assume that x is infinite. Show that the family $F = \{y \subseteq x : x \backslash y \text{ is finite}\}$ is a proper filter on x.

Subsets of a set x that have finite complements are often called *cofinite subsets of x,* and the filter F described in Exercise 12(e) is often called the *cofinite filter on x.*

As the examples in parts (d) and (e) of the above exercise indicate, filters are collections of "large" subsets of a given set x, where the notion of "largeness" depends on the particular filter.

A filter F on x is called an *ultrafilter* if F is proper, and for every filter G on x which satisfies $F \subseteq G$, either $G = F$ or $G = \mathcal{P}(x)$. In other words, an ultrafilter is a maximal proper filter. The following theorem gives a very useful characterization of ultrafilters.

Theorem 10: *Let $x \neq \emptyset$, and let F be a filter on x. Then F is an ultrafilter on x iff for all $y \subseteq x$ exactly one of the sets y, $x \backslash y$ is in F.*

Proof: Let x, F be as in the assumptions. The "if"-direction is easy: Assume that for each $y \subseteq x$ exactly one of the sets y, $x \backslash y$ is in F. Since exactly one of the sets $x, \emptyset$ is in F, and since $x \in F$ by Exercise 12(a), it follows that $\emptyset \notin F$. Thus F is proper. Moreover, if G is a filter on x that contains F and if $y \in G \backslash F$, then $x \backslash y \in F$. Thus $\emptyset = y \cap (x \backslash y) \in G$. By Exercise 12(b), $G = \mathcal{P}(x)$.

Let us now prove the contrapositive of the "only if"-direction. Assume $y \subseteq x$ is such that $y \notin F$ and $x \backslash y \notin F$. We want to construct a proper filter $G \neq F$ that contains F, thus contradicting maximality of F.

Lemma 11: *Let F be a filter on a nonempty set x, and let $y \subseteq x$. Define $flt(F, y) = \{w \subseteq x : \exists z \in F\,(y \cap z \subseteq w)\}$. Then:*

(a) $flt(F, y)$ is a filter on x that contains $F \cup \{y\}$;

(b) $flt(F, y)$ is proper iff $x \backslash y \notin F$;

(c) $flt(F, y) \neq F$ iff $y \notin F$.

EXERCISE 13(G): (a) Derive Theorem 10 from Lemma 11.

(b) Prove Lemma 11. □

EXERCISE 14(G): Let x be a nonempty set, and let $a \in x$. Show that the set $F_a = \{y \subseteq x : a \in y\}$ is an ultrafilter on x. *Hint:* For showing that F_a is a filter, you may want to use a previous exercise.

An ultrafilter F on a set x is called a *principal ultrafilter* if $F = F_a$ for some $a \in x$. Principal ultrafilters are also said to be *fixed,* and nonprincipal ones are also called *free ultrafilters.*

EXERCISE 15(PG): Show that if x is a finite nonempty set and F is an ultrafilter on x, then $F = F_a$ for some $a \in x$.

The question arises whether there are any nonprincipal ultrafilters whatsoever. Exercise 15 shows that every set which carries a nonprincipal ultrafilter[4] must be infinite. Moreover, it follows from the same exercise that a nonprincipal ultrafilter cannot contain a finite set.

EXERCISE 16(G): Why?

In other words, if H is a nonprincipal ultrafilter on any set x, then H contains the filter of cofinite subsets of x described in Exercise 12(e). It also works the other way round: If H is an ultrafilter on a set x such that every cofinite subset of x is in H, then H is nonprincipal. Thus, if we knew that the filter of cofinite subsets of a set x can be extended to an ultrafilter, then we could conclude that there exists a nonprincipal ultrafilter on x. Zorn's Lemma provides the missing link.

Theorem 12: *Let x be a nonempty set, and let F be a proper filter on x. Then there exists an ultrafilter H on x such that $F \subseteq H$.*

Proof: Let x and F be as in the assumptions. This proof will be our prototype for an application of Zorn's Lemma. Therefore, our narrative will oscillate between two levels of abstraction: While describing how to handle the specific problem we are dealing with, we shall simultaneously discuss the technique of applying Zorn's Lemma in general.

The crucial step in any proof using Zorn's Lemma is finding a p.o. to which the lemma can be meaningfully applied. What kind of p.o. could serve our purpose? We want an ultrafilter H on x such that $F \subseteq H$. All we can get from Zorn's Lemma is a maximal element in some p.o. Therefore, we must find a p.o. $\langle \mathcal{X}, \preceq \rangle$ such that, given a maximal element of it, we can easily construct an ultrafilter with the desired properties. Ideally, a maximal element would just *be* an ultrafilter H on x such that $F \subseteq H$. What kind of objects should the other elements of $\mathcal{X}$ be? Some sort of approximations to H. In our case, the most natural candidate for $\mathcal{X}$ is the set of all proper filters G on x such that $F \subseteq G$. The ultrafilter H we are seeking is by definition an element of $\mathcal{X}$ that is maximal with respect to inclusion. Thus, $\subseteq$ is the most natural choice for the partial order relation $\preceq$.

Having chosen the p.o. $\langle X, \preceq \rangle$, and having convinced ourselves that any maximal element in this p.o. really gives us what we want, it only remains to verify that $\langle \mathcal{X}, \preceq \rangle$ satisfies the assumptions of (ZL). This part of the proof is usually straightforward, although it can be tedious.

First, one should verify that the relation $\preceq$ is indeed a partial order relation. Since in our case $\preceq$ is just inclusion, we get this part of the argument for free as a consequence of Exercise 2.2(c).

Second, we need to verify that every chain in $\langle \mathcal{X}, \preceq \rangle$ has an upper bound. So, let $\mathcal{C}$ be a chain in $\mathcal{X}$. We want to find an upper bound for $\mathcal{C}$. One needs to distinguish two cases.

Case 1: $\mathcal{C} = \emptyset$. Then every element of $\mathcal{X}$ is an upper bound of $\mathcal{C}$; so this case boils down to showing that $\mathcal{X}$ is nonempty. In the present proof, for this it suffices to observe that $F \in \mathcal{X}$.[5]

[4] This is a convenient way of saying that there exists a nonprincipal ultrafilter on this set.

[5] Although the case $\mathcal{C} = \emptyset$ may appear to have no substance at all, it cannot be ignored.

Case 2: $\mathcal{C} \neq \emptyset$. The proof splits into two parts: guessing a candidate for an upper bound C of $\mathcal{C}$, and verifying that this candidate C is indeed an upper bound for $\mathcal{C}$. Guessing an upper bound for $\mathcal{C}$ may occasionally require some creativity, but in cases where $\preceq$ is inclusion, taking $C = \bigcup \mathcal{C}$ usually works.

So let us conclude the proof of Theorem 12 by showing that $C = \bigcup \mathcal{C}$ is an upper bound of $\mathcal{C}$. Since $\bigcup \mathcal{C}$ is a superset of every element of $\mathcal{C}$, the only thing we really need to check is that $C \in \mathcal{X}$. We have to verify four properties:

(1) $F \subseteq C$: This follows from the fact that $F \subseteq G \subseteq \bigcup \mathcal{C}$ for every $G \in \mathcal{C}$.

(2) C is closed under supersets: Let $y \in C$, and let z be such that $y \subseteq z \subseteq x$. Then there exists some $G \in \mathcal{C}$ such that $y \in G$. Since G is closed under supersets, $z \in G$. But then z is also in $\bigcup \mathcal{C}$.

(3) C is closed under finite intersections: Let $y_0, \dots, y_k \in C$, and let $G_0, \dots, G_k$ be filters in $\mathcal{C}$ such that $y_i \in G_i$ for each $i \leq k$. Now we are going to use the fact that $\mathcal{C}$ is a chain. Since $\mathcal{C}$ is linearly ordered by inclusion, we may without loss of generality assume that $G_i \subseteq G_{i+1}$ for all $i < k$. Then $\{y_0, \dots, y_k\} \subseteq G_k$, and since G_k is closed under finite intersections, $y_0 \cap \dots \cap y_k \in G_k$. Since $G_k \subseteq C$, we are done.

(4) C is proper: Suppose not. Then $\emptyset \in C$, and hence $\emptyset \in G$ for some $G \in \mathcal{C}$. But this is impossible, since every element of $\mathcal{C}$ was supposed to be a proper filter.

We have proved Theorem 12. □

EXERCISE 17(G): At which point of the argument in Case 2 did we use the assumption that $\mathcal{C} \neq \emptyset$?

EXERCISE 18(PG): A filter F on a set X is *closed under countable intersections* or *countably complete* if it satisfies the condition:

$$(\text{ii}^+) \qquad \forall H \subseteq F\,(H \text{ is countable } \rightarrow \bigcap H \in F).$$

(a) Show that every principal ultrafilter is countably complete.

(b) Let x be an uncountable set. Show that the *cocountable filter on* x, i.e., the family $F = \{y \subseteq x : x \backslash y \text{ is countable}\}$, is a countably complete filter on x.

(c) It is not true that for every uncountable set x the cocountable filter on x can be extended to a countably complete ultrafilter. Therefore, if one replaces all occurrences of the word "(ultra)filter" in Theorem 12 and its proof by "countably complete (ultra)filter," then the argument must break down at some point. Where is this point?

For later reference, let us assign a number to the following important consequence of Theorem 12.

Corollary 13: *Let x be an infinite set. Then there exists a nonprincipal ultrafilter on x.*

EXERCISE 19(G): Use the method of the proof of Theorem 12 to show that every vector space has a basis.

Let a, b be infinite subsets of ω. We say that a and b are *almost disjoint* if $a \cap b$ is finite. For example, every two disjoint sets are almost disjoint; and the sets of prime numbers and even numbers are almost disjoint, but not disjoint. We say that a family A of infinite subsets of ω is a *mad family* (a maximal almost disjoint family) if every two elements of A are almost disjoint, and, for every infinite subset $x \subseteq \omega$, there exists an $a \in A$ such that $x \cap a$ is infinite.

EXERCISE 20(G): Use the method of the proof of Theorem 12 to show that every family of infinite, pairwise almost disjoint subsets of ω is contained in a mad family.

EXERCISE 21(PG): Use the method of the proof of Theorem 12 to show that (ZL) implies (HMP). *Hint:* You will be working with chains in two different p.o.'s. Be careful to make notational distinctions.

Our next theorem closes the loop and shows that all of the following are equivalent in ZF: (A9), (AC), (DIS), (WO), (ZL), (HMP).

Theorem 14(ZF): (ZL) $\rightarrow$ (WO).

Proof: Let X be an arbitrary set. We want to find a wellorder relation $\trianglelefteq$ on X. The definition of a wellorder relation on X does not mention any kind of maximality. How could Zorn's Lemma possibly be of any help? Recall that the partially ordered set $\mathcal{X}$ in Zorn's Lemma should consist of some kind of approximations to the object we want to construct. Thus, a natural candidate for $\mathcal{X}$ would be the set of all pairs $\langle Y, \trianglelefteq\rangle$, where $Y \subseteq X$ and $\trianglelefteq$ is a wellorder relation on Y. This set is always nonempty, even if $X = \emptyset$.

EXERCISE 22(G): Why?

Can we define a partial order relation $\preceq$ on $\mathcal{X}$ such that each $\preceq$-maximal element is of the form $\langle X, \trianglelefteq\rangle$?

Claim 15: *Let $Y \subset X$, let $\trianglelefteq_0$ be a wellorder relation on Y, and let $x \in X \backslash Y$. Then the relation $\trianglelefteq_1 = \trianglelefteq_0 \cup \{\langle y, x\rangle : y \in Y \cup \{x\}\}$ is a wellorder relation on $Y \cup \{x\}$.*

EXERCISE 23(G): Prove Claim 15.

It follows from the claim that if we define:

$$\langle Y, \trianglelefteq_Y\rangle \preceq \langle Z, \trianglelefteq_Z\rangle \text{ iff } Y \subseteq Z \ \& \ \trianglelefteq_Y \subseteq \trianglelefteq_Z,$$

then $\langle Y, \trianglelefteq\rangle$ is maximal in $\langle \mathcal{X}, \preceq\rangle$ if and only if $Y = X$.

This starts to look promising; let us see whether $\langle \mathcal{X}, \preceq\rangle$ satisfies the assumptions of Zorn's Lemma. By Exercises 2.2(c) and 2.26(a), $\preceq$ is a partial order relation. We need to check whether every chain has an upper bound. Let $\mathcal{C}$ be a subset of $\mathcal{X}$ that is linearly ordered by $\preceq$. A natural candidate for an upper bound of $\mathcal{C}$ is the pair $C = \langle C_0, C_1\rangle$, where $C_0 = \bigcup\{Y : \exists \trianglelefteq \ (\langle Y, \trianglelefteq\rangle \in \mathcal{C})\}$ and $C_1 = \bigcup\{\trianglelefteq : \exists Y \ (\langle Y, \trianglelefteq\rangle \in \mathcal{C})\}$. We want to show that $C \in \mathcal{X}$.

EXERCISE 24(G): Prove that $C_0 \subseteq X$ and that C_1 is a linear order relation on C_0.

EXERCISE 25(PG): Is C_1 always a wellfounded relation?

In general, when working on the proof of a new theorem, you should split your efforts between trying to prove the theorem and trying to construct a counterexample.[6] Since we are unable to prove wellfoundedness of C_1 (or did you succeed?), let us see whether we can find a counterexample. By Theorem 2.2 it suffices to find a $\preceq$-chain $\mathcal{C}$ such that, if C is defined as above, then there are $x_0, x_1, \dots, x_k, \dots \in C_0$ with $\langle x_{i+1}, x_i\rangle \in C_1$ for each $i \in \omega$.

Having clearly formulated what kind of counterexample one wants, it is often possible to follow the path of least resistence and let the counterexample construct itself. In our case, we need a chain of w.o.'s. Of course no finite chain will do.

EXERCISE 26(G): Why?

So let us try the simplest possible kind of infinite chain: $\mathcal{C} = \{\langle Y_k, \trianglelefteq_k\rangle : k \in \omega\}$, where $\langle Y_k, \trianglelefteq_k\rangle \preceq \langle Y_{k+1}, \trianglelefteq_{k+1}\rangle$ for all $k \in \omega$. We also need $\{x_0, x_1, \dots, x_k, \dots\} \subseteq \bigcup_{k\in\omega} Y_k$. Why don't we just fix a set $\{x_k : k \in \omega\}$, and let $Y_k = \{x_0, \dots, x_{k+1}\}$? Moreover, if $C_1 = \bigcup_{k\in\omega} \trianglelefteq_k$, then we should have $\langle x_{i+1}, x_i\rangle \in C_1$ for all $i \in \omega$. This implies that we have no other choice but to put $\trianglelefteq_k = \{\langle x_i, x_j\rangle : j \leq i \leq k\}$ for all $k \in \omega$.

Did we exaggerate when we claimed that the counterexample would construct itself? Well, we still should double-check that our construction indeed yields a counterexample.

EXERCISE 27(G): Show that $\mathcal{C} = \{\langle Y_k, \trianglelefteq_k\rangle : k \in \omega\}$ is a chain in $\langle \mathcal{X}, \preceq\rangle$ for which the corresponding pair $\langle C_0, C_1\rangle = \langle \bigcup_{k\in\omega} Y_k, \bigcup_{k\in\omega} \trianglelefteq_k\rangle$ is not a w.o.

Our proof of Theorem 14 has run into a dead-end street. What can one do in such a predicament? Basically, there is a choice between three courses of action. The worst of these is giving up. The second best is collecting a consolation prize in the form of a partial result.

EXERCISE 28(G): Prove that every set can be linearly ordered. *Hint:* Slightly modify the set $\mathcal{X}$ and verify that the argument up to and including Exercise 24 goes through for this modified version of $\mathcal{X}$.

The best course of action is backing off a little and modifying the proof so that it will yield the desired result. This is not always possible though, and sometimes a partial result is all one can get.

If one chooses the third course of action, the question arises: "How far should I back off?" Only as far as is necessary to eliminate the troublesome counterexample.[7] In our case, the smallest imaginable concession would be leaving the definition of $\langle \mathcal{X}, \preceq\rangle$ unchanged and finding a new candidate for an upper bound of a chain $\mathcal{C}$. If this were possible, we would not have to waste the effort that went into checking that $\preceq$ is a partial order relation and that $\preceq$-maximal elements are w.o.'s of X. Unfortunately, we cannot get off the hook so easily.

[6] Even the most experienced mathematicians occasionally forget this simple piece of common sense and waste their time and energy on trying to prove an "obvious" (but false) theorem without even considering the possibility that there might be a counterexample.

[7] In a real search for a proof, you may have to go repeatedly through the cycle of constructing counterexamples, backing off, and exploring a new direction. As long as there is room to back off and go into a hitherto unexplored direction, you should not give up. But if you get the distinctive feeling of being caught in an infinitely repeating loop, then it *is* time to give up or at least to gain some distance from the problem.

EXERCISE 29(G): Show that the chain $\mathcal{C} = \{\langle Y_k, \trianglelefteq_k\rangle : k \in \omega\}$ of Exercise 27 does not have *any* upper bound in $\langle \mathcal{X}, \preceq\rangle$.

Too bad. So we have to modify the definition of $\langle \mathcal{X}, \preceq\rangle$. As a matter of parsimony, let us try to leave $\mathcal{X}$ as is and only replace $\preceq$ by another relation $\preceq_e$.[8] We will have to choose $\preceq_e$ in such a way that every maximal element of $\langle \mathcal{X}, \preceq_e\rangle$ is a w.o. of X. Claim 15 gave us the analogous property for $\preceq$; could we perhaps choose $\preceq_e$ in such a way that Claim 15 would remain relevant?

The claim says that if we stick an extra element on top of a w.o. $\langle Y, \trianglelefteq\rangle$, then we again obtain a w.o. In our counterexample, the new element x_{k+1} was always added *below* each element in Y_k, and this is exactly what got us into trouble. So, why don't we define: $\langle Y, \trianglelefteq_Y\rangle \preceq_e \langle Z, \trianglelefteq_Z\rangle$ iff $Y \subseteq Z$ & $\trianglelefteq_Y \subseteq \trianglelefteq_Z$ & $\forall z \in Z \backslash Y \, \forall y \in Y \, (\langle y, z\rangle \in \trianglelefteq_Z)$?

In other words, $\langle Y, \trianglelefteq_Y\rangle \preceq_e \langle Z, \trianglelefteq_Z\rangle$ iff the w.o. $\langle Z, \trianglelefteq_Z\rangle$ is an *end-extension* (see the definiton of end-extension in Chapter 2) of the w.o. $\langle Y, \trianglelefteq_Y\rangle$.

We can still deduce from Claim 15 that every $\preceq_e$-maximal element is a w.o. of the form $\langle X, \trianglelefteq\rangle$. Let us check whether $\langle \mathcal{X}, \preceq_e\rangle$ satisfies the assumptions of Zorn's Lemma.

EXERCISE 30(G): Show that $\preceq_e$ is a partial order relation on $\mathcal{X}$.

Let $\mathcal{C}$ be a chain in $\langle \mathcal{X}, \preceq_e\rangle$, and let $C = \langle C_0, C_1\rangle$ be defined exactly as in our previous attempt: $C_0 = \bigcup\{Y : \exists \trianglelefteq \, (\langle Y, \trianglelefteq\rangle \in \mathcal{C})\}$ and $C_1 = \bigcup\{\trianglelefteq : \exists Y \, (\langle Y, \trianglelefteq\rangle \in \mathcal{C})\}$. Since $\mathcal{C}$ is also a chain in $\langle \mathcal{X}, \preceq\rangle$, we can conclude from Exercise 24 that C_1 is a linear order relation on C_0.

EXERCISE 31(G): Show that C is an end-extension of every $\langle Y, \trianglelefteq\rangle \in \mathcal{C}$; i.e., show that if $y \in Y$ and $x \in C_0 \backslash Y$, then $\langle y, x\rangle \in C_1$.

Now let us verify that we have succeeded in eliminating all counterexamples to wellfoundedness.

Let C_2 be a nonempty subset of C_0. We want to show that C_2 has a C_1-minimal element. Fix $\langle Y, \trianglelefteq\rangle \in \mathcal{C}$ such that $Y \cap C_2 \neq \emptyset$. Let y be the $\trianglelefteq$-minimum element of $C_2 \cap Y$. If $x \in C_2$, then either $x \in Y$, and hence $\langle y, x\rangle \in C_1$ by minimality of y in Y, or $x \in C_0 \backslash Y$, and $\langle y, x\rangle \in C_1$ by Exercise 31. This shows that y is the C_1-minimum element of C_2. Since C_2 was an arbitrary nonempty subset of C_0, we have shown that C_1 is a wellorder relation on C_0. Therefore, $C \in \mathcal{X}$, and we have proved Theorem 14. □

You will see many applications of (WO) later in this book. For now, we'll show you just one. The following theorem answers a question we pondered in Chapter 2, right after Exercise 2.15.

Theorem 16: *Let X be an arbitrary set, and let W be a wellfounded relation on X. Then there exists a wellorder relation R on X such that $W \subseteq R$.*

EXERCISE 32(G): Show that (WO) is a special case of Theorem 16. Conclude that Theorem 16 is another statement that is equivalent in ZF to (AC).

Proof of Theorem 16: Let X be given, and let W be a wellfounded relation on X. By (WO), there exists a wellorder relation on $\mathcal{P}(X)$. Fix such a relation $\preceq$. By Theorem 3.5, $|X| < |\mathcal{P}(X)|$. By Theorem 4.34, there exists a rank function

[8] Trying to restrict $\preceq$ to a subset of $\mathcal{X}$ would be an even more parsimonious approach, but it does not work here.

$rk : X \to \mathcal{P}(X)$ for W with respect to $\preceq$. For each $\alpha \in rng(rk)$, let $\mathcal{W}_\alpha$ be the set of all wellorder relations on $rk^{-1}\{\alpha\}$. By (WO), for each $\alpha \in rng(rk)$ the set $\mathcal{W}_\alpha$ is nonempty. Now it follows from (AC) that there exists some $f \in \prod_{\alpha \in rng(rk)} \mathcal{W}_\alpha$. Fix such an f, and define for $x, y \in X$: $\langle x, y\rangle \in R$ iff $rk(x) < rk(y)$ or $rk(x) = rk(y) \wedge \langle x, y\rangle \in f(rk(x))$.

EXERCISE 33(PG): Prove that R is a wellorder relation on X. □

The versions of the Axiom of Choice discussed so far are all set-theoretic in nature. But there are also statements equivalent to (AC) that "belong" to other branches of mathematics. Here is the most prominent example from general topology, known as Tychonoff's Theorem:

(TY) The product of every family of compact spaces is compact.

The proof of Tychonoff's Theorem in ZFC will be given in Chapter 13. Here we show the following:

Theorem 17 (ZF): (TY) $\to$ (AC).

Proof: Let $\{A_i : i \in I\}$ be an indexed family of nonempty sets. We show that there exists a choice function $f \in \prod_{i \in I} A_i$. Let $*$ be an object that does not belong to any of the A_i's, and let $X_i = A_i \cup \{*\}$ for every $i \in I$. If $\tau_i = \{\emptyset, \{*\}, X_i\}$ for each $i \in I$, then τ_i is a compact topology on X_i, and (TY) implies that $\prod_{i \in I} X_i$ with the product topology is also compact. Now consider the family $\mathcal{U} = \{U_i : i \in I\}$, where $U_j = \{f \in \prod_{i \in I} A_i : f(j) = *\}$. Since $*$ is an isolated point in X_j, each of the sets U_j is open. Moreover, if J is a finite subset of I, then we do not need the Axiom of Choice to construct an $f \in \prod_{i \in I} X_i \backslash \bigcup_{j \in J} U_j$: Just hand-pick $f(j) \in A_j$ for the finitely many $j \in J$, and let $f(i) = *$ for $i \in I \backslash J$. This means that no finite subfamily of $\mathcal{U}$ is a cover of $\prod_{i \in I} X_i$. Thus by compactness of $\prod_{i \in I} X_i$, the family $\mathcal{U}$ is not a cover of $\prod_{i \in I} X_i$. In other words, there exists $f \in \prod_{i \in I} X_i \backslash \bigcup \mathcal{U}$. This f must be an element of $\prod_{i \in I} A_i$; and this is exactly what we want. □

EXERCISE 34(PG): The spaces $\langle X_i, \tau_i\rangle$ constructed in the proof of Theorem 17 are not Hausdorff. It is known that the statement "The product of every family of compact Hausdorff spaces is compact" is not equivalent to the Axiom of Choice.[9] Given this information, show that the following is not provable in ZF: "For every indexed family $\{X_i : i \in I\}$ of nonempty sets, there exists a family $\{\tau_i : i \in I\}$ such that $\langle X_i, \tau_i\rangle$ is a compact Hausdorff space for every $i \in I$."

Not surprisingly, the Axiom of Choice plays a prominent role in the study of cardinals. Theorem 3.4 asserts that the following holds for arbitrary nonempty sets A, B: There exists an injection $f : A \to B$ if and only if there exists a surjection $g : B \to A$. The "only if"–direction is a theorem of ZF, but the "if"–direction requires the Axiom of Choice.

EXERCISE 35(G): Dig out your solution to Exercise 3.5, and convince yourself that the "only if"–direction is a theorem of ZF. Find the step in the proof of the "if"–direction where a version of the Axiom of Choice was used.

Let us put down for later reference which half of Theorem 3.4 can be proved in ZF.

[9]The proof of Tychonoff's Theorem, both for Hausdorff spaces and for arbitrary topological spaces, will be given in Chapter 13 of Volume II.

Fact 18 (ZF): *Let A, B be nonempty sets such that there exists an injection from A into B. Then there exists a surjection from B onto A.*

The next theorem gives a sample of further properties of cardinalities that are equivalent to the Axiom of Choice. The symbol $[X]^{<\aleph_0}$ stands for the family of finite subsets of a set X.

Theorem 19 (ZF): *The following statements are all equivalent to* (AC):

(IC) *For arbitrary sets X, Y,*
either there exists an injection from X into Y
or there exists an injection from Y into X.

(SC) *For arbitrary nonempty sets X, Y,*
either there exists a surjection from X onto Y
or there exists a surjection from Y onto X.

(FS) $[X]^{<\aleph_0} \approx X$ *for every infinite set X.*

The symbols in parantheses stand for "injective comparability," "surjective comparability," and "finite subsets."

EXERCISE 36(G): Show in ZF that (WO) implies (IC), and that (IC) implies (SC).

Experienced mathematicians will not be particularly surprised by the equivalences we have presented so far. For good measure, let us give a more bizarre example.

Theorem 20 (ZF): *The following is equivalent to* (AC):

(AG) *For every nonempty set X there exist an $e \in X$ and a binary function $* : X \times X \to X$ such that the structure $\langle X, *, e\rangle$ is an abelian group.*

In this chapter we prove only one half of Theorem 20. The proofs of the other half of Theorem 20 and of the remaining parts of Theorem 19 will be given in installments spaced over the next two chapters (see Lemma 10.17, Exercise 11.3(b), and Exercise 11.22).

Lemma 21 (ZF): (FS) $\to$ (AG)

Proof: Let X be any nonempty set. Note that if there is any abelian group $\langle Y, \oplus, \mathbf{1}\rangle$ such that $Y \approx X$, then it is easy to define $e \in X$ and an operation $* : X^2 \to X$ as required. Now consider two cases:

- If X is finite, then $X \approx Z_n$ for some $n \in \omega$, where Z_n is the cyclic group of order n.
- If X is infinite, then by (FS), $X \approx [X]^{<\aleph_0}$, and the following claim implies the lemma.

Claim 22 (ZF): *For every nonempty set X, the structure $\langle [X]^{<\aleph_0}, \Delta, \emptyset\rangle$ is an abelian group.*

EXERCISE 37(G): Prove Claim 22. □

One might ask whether the Axiom of Choice also holds for proper classes. For example, consider the following statement:

(AC)$^+$ There exists a functional class $\mathbf{F}$ with domain $\mathbf{V}\backslash\{\emptyset\}$ such that $\mathbf{F}(x) \in x$ for every $x \in \mathbf{V}\backslash\{\emptyset\}$.

If $\mathbf{F}$ is as above, then there exists a formula φ of L_S with two free variables and possibly some parameters such that $\langle x, y\rangle \in \mathbf{F}$ if and only if $\varphi(x, y)$ holds. Thus, $(\mathrm{AC})^+$ asserts that there exists a *definable* choice function for the class of all nonempty sets. This is a rather strong assertion, and it should not be surprising that $(\mathrm{AC})^+$ is not a theorem of ZFC. However, if ZF is a consistent theory, then so is the theory ZF + $(\mathrm{AC})^+$ (see Corollary 12.33).

9.2. Set Theory without the Axiom of Choice

Let us return to the question: What is so special about the Axiom of Choice that it contributes one third of the letters in "ZFC"? In Chapter 7 we tried to illustrate by means of an analogy that there is something fishy about the idea that one should always be able to make infinitely many choices at the same time. But is making infinitely many choices really more implausible than forming the power set of any given set, or, for that matter, treating infinite collections as single entities called sets?

A deeper reason why the Axiom of Choice has attracted more skepticism than the other axioms of ZFC is this: There are some extremely counterintuitive theorems of ZFC that are not provable in ZF alone. The most notorious of these is perhaps the famous *Banach-Tarski Paradox*. It states that if B is any (open or closed) ball in $\mathbb{R}^3$, then there exists a partition of B into pairwise disjoint pieces $B_0, \ldots, B_4$ such that B_0, B_1, B_2 can be reassembled to form the whole ball B, and the same is true for B_3 and B_4! Speaking more precisely, there are isometries $\varphi_0, \ldots, \varphi_4 : \mathbb{R}^3 \to \mathbb{R}^3$ such that

$$B = \varphi[B_0] \cup \varphi[B_1] \cup \varphi[B_2] = \varphi[B_3] \cup \varphi[B_4].$$

Now, if this were true, would Banach and Tarski have published their result? Or would they rather have gone into the business of cutting up balls of solid gold into pieces and reassembling these pieces into two identical copies of each of the original balls?[10] Obviously, it must be impossible to base a technological process on the Banach-Tarski Paradox.

But why is this so obvious? Well, in real life, every solid ball B has a certain mass $m(B) > 0$. If we cut it up into pairwise disjoint pieces $B_0, \ldots, B_4$, then $m(B) = \sum_{i=0}^{4} m(B_i)$. Moreover, moving around a piece doesn't change its mass. In other words, if φ_i is an isometry of $\mathbb{R}^3$, then $m(\varphi[B_i]) = m(B_i)$. Thus, $\sum_{i=0}^{4} m(\varphi[B_i]) = \sum_{i=0}^{4} m(B_i) = m(B)$. On the other hand, $\sum_{i=0}^{4} m(\varphi[B_i]) = \sum_{i=0}^{2} m(\varphi[B_i]) + \sum_{i=3}^{4} m(\varphi[B_i]) = m(B) + m(B) = 2m(B)$, which is clearly a contradiction.

There is a gap in the above argument though. How do we know that each of the pieces has a mass? Of course, if you take a real ball of solid gold and cut it into pieces, then each piece will have a mass, and the masses of the pieces will add up to the masses of the original ball. But each real ball is a collection of finitely many atoms, and the mass of (each piece of) the ball is the sum of the masses of all atoms that belong to (this piece of) the ball. In contrast, in the proof of the Banach-Tarski Paradox, each of the pieces B_i is assembled of individual points of

---

[10] In case you would like to take up the opportunity they missed, we give a proof of the Banach-Tarski Paradox in Section 9.4. Strictly speaking, Banach and Tarski only showed that a paradoxical decomposition of the ball into finitely many pieces exists. It was shown later by R. Robinson that five pieces suffice. S. Wagon's book *The Banach-Tarski Paradox*, Cambridge University Press, 1993, gives a comprehensive treatment of topics related to this paradox.

$\mathbb{R}^3$ in a somewhat haphazard way (using the Axiom of Choice). Since points in $\mathbb{R}^3$ are infinitesimally small, each individual point is supposedly massless. Somewhat mysteriously, the massless points contained in the ball B add up to a set with finite, nonzero mass. But does this imply that we should be able to assign a mass to just *any* collection of points in B?

Let us reformulate this question in mathematical terms. Suppose $\mathcal{M}$ is a family of subsets of a set X, and let $m : \mathcal{M} \to [0, +\infty]$. We call m a *measure* if

(1) $m(\emptyset) = 0$ and

(2) $m(\bigcup_{n \in \omega} A_n) = \sum_{n \in \omega} m(A_n)$ for every countable family $\{A_n : n \in \omega\}$ of pairwise disjoint elements of $\mathcal{M}$.

Of course, this definition makes sense only if $\emptyset \in \mathcal{M}$, and if $\mathcal{M}$ is closed under countable unions of pairwise disjoint sets, that is, if for every countable family $\{A_n : n \in \omega\}$ of pairwise disjoint elements of $\mathcal{M}$ also $\bigcup_{n \in \omega} A_n \in \mathcal{M}$. If $\mathcal{M}$ is also closed under intersections and complements, i.e., if for all $A, B \in \mathcal{M}$ the sets $A \cap B$ and $X \backslash A$ are in $\mathcal{M}$, then $\mathcal{M}$ is called a *σ-field* or a *σ-algebra* of subsets of X.

EXERCISE 38(G): (a) Show that if $\mathcal{M}$ is a σ-algebra, then $\mathcal{M}$ is closed under arbitrary countable unions, i.e., show that if $\mathcal{A}$ is a countable family of (not necessarily pairwise disjoint) elements of $\mathcal{M}$, then $\bigcup \mathcal{A} \in \mathcal{M}$.

(b) Show that if m is a measure whose domain is a σ-algebra $\mathcal{M}$ and if $A, B \in \mathcal{M}$ are such that $A \subseteq B$, then $m(A) \leq m(B)$.

It is customary to require in the definition of a measure m that its domain $\mathcal{M}$ is a σ-algebra, and we shall henceforth make this requirement. $\mathcal{M}$ is called the σ-algebra of *m-measurable* subsets of X.

We say that a measure m *vanishes on the points* if $\{x\} \in \mathcal{M}$ and $m(\{x\}) = 0$ for all $x \in X$. Given a group G of bijections from X into X, we say that m is *G-invariant* if for all $A \in \mathcal{M}$ and $g \in G$, $g[A] \in \mathcal{M}$ and $m(g[A]) = m(A)$.

The mathematical equivalent of the mass of a three-dimensional object is the notion of three-dimensional *Lebesgue measure*. This is a measure m_3 defined on a certain σ-algebra $\mathcal{L}_3$ of subsets of $\mathbb{R}^3$. Three-dimensional Lebesgue measure vanishes on the points, is invariant under isometries of $\mathbb{R}^3$, and assigns to each ball in $\mathbb{R}^3$ its volume. Therefore, if $\{B_0, \ldots, B_4\}$ is a decomposition of a ball B as in the Banach-Tarski Paradox, then not all the pieces can be Lebesgue-measurable (i.e., in $\mathcal{L}_3$). Moreover, it follows from this paradox that there exists no measure $m : \mathcal{P}(\mathbb{R}^3) \to [0, +\infty]$ whatsoever such that $m_3 = m|\mathcal{L}_3$ and m is invariant under isometries of $\mathbb{R}^3$. This conclusion from the Banach-Tarski Paradox is certainly somewhat counterintuitive; but as far as we know, it does not lead to a mathematical contradiction.

To show you how the Axiom of Choice is used in the formation of nonmeasurable sets, let us now present one such construction.

One-dimensional Lebesgue measure is a certain measure m_1 that is defined on a σ-algebra $\mathcal{L}_1$ of subsets of $\mathbb{R}$. This measure vanishes on the points, is invariant under isometries of $\mathbb{R}$ (in particular, m_1 is *translation-invariant*), and assigns to each interval in $\mathbb{R}$ its length. A measure m is said to *extend* m_1 if m is defined on some σ-algebra of subsets of $\mathbb{R}$ and $m|\mathcal{L}_1 = m_1$. A set $V \subset \mathbb{R}$ is called a *Vitali set* if V is not m-measurable for any translation-invariant measure m that extends m_1.

Theorem 23: *There exists a Vitali set.*

Proof: On $[0,1)$, define an equivalence relation $\sim$ by: $x \sim y$ iff $x - y \in \mathbb{Q}$. By the Axiom of Choice, there exists a selector V for $\sim$, i.e., a set $V \subset [0,1)$ such that $|V \cap x/_{\sim}| = 1$ for every equivalence class $x/_{\sim}$.

We show that V is as required. For every $q \in \mathbb{Q}$, consider the set $q+V = \{q+x : x \in V\}$.

EXERCISE 39(G): Show that if $q, r \in \mathbb{Q}$ and $q \neq r$, then $(q+V) \cap (r+V) = \emptyset$.

Now let m be a translation-invariant measure that extends m_1. Then $m(q+V) = m(V)$ for every $q \in \mathbb{Q}$. Therefore, if $W = \bigcup_{q \in [-1,1] \cap \mathbb{Q}} (q+V)$, then

$$m(W) = \sum_{q \in [-1,1] \cap \mathbb{Q}} m(q+V) = \sum_{q \in [-1,1] \cap \mathbb{Q}} m(V).$$

Thus, if $m(V) = 0$, then $m(W) = 0$; and if $m(V) > 0$, then $m(W) = +\infty$. But $[0,1) \subseteq W \subseteq [-1,2)$, and hence

$$1 = m_1([0,1)) = m([0,1)) \leq m(W) \leq m([-1,2)) = m_1([-1,2)) = 3,$$

a contradiction. □

In particular, Theorem 23 implies that not all subsets of the real line are Lebesgue measurable. Now the question arises: Can one-dimensional Lebesgue measure be extended to *any* (not necessarily translation-invariant) measure defined on all subsets of $\mathbb{R}$? We shall return to this intriguing question in Chapter 23.

Let us briefly mention another important construction of nonmeasurable sets.

Definition 24: Let X be a topological space. A set $B \subset X$ is called a *Bernstein set* if $F \cap B \neq \emptyset$ and $F \backslash B \neq \emptyset$ for every uncountable closed set $F \subset X$.

Theorem 11.4 shows that for every positive integer n there is a Bernstein set $B \subset \mathbb{R}^n$.

Lebesgue measure (of any dimension n) is *regular,* which means that if $A \subset \mathbb{R}^n$ is Lebesgue-measurable, then $m_n(A) = \sup\{m_n(K) : K \subseteq A \wedge K \text{ is closed}\} = \inf\{m_n(G) : A \subseteq G \wedge G \text{ is open}\}$.

It follows that if B is a Bernstein set in $\mathbb{R}^n$, then B is not Lebesgue measurable. To see this, note that if K is a closed subset of B, then K is countable, hence $m_n(B) = 0$ (recall that m_n vanishes on the points). Similarly, if G is an open superset of B, then $\mathbb{R}^n \backslash G$ is countable, hence of measure zero. Thus, $\sup\{m_n(K) : K \subseteq A \wedge K \text{ is closed}\} = 0 < \inf\{m_n(G) : A \subseteq G \wedge G \text{ is open}\} = +\infty$.

It can be shown that if ZF is consistent, then so is the theory ZF + "every subset of the real line is Lebesgue measurable." Since the existence of nonmeasurable sets of reals is somewhat counterintuitive, it makes sense to try to develop mathematics without the Axiom of Choice. As it turns out, many useful mathematical theorems are indeed provable in ZF alone. So the questions arises: Why do most mathematicians accept the Axiom of Choice?

In our opinion, there are three reasons for this:

Reason 1: The Axiom of Choice is necessary for the development of many mathematical theories.

Reason 2: In many cases, even if a theorem can be proved in ZF, a clever use of (AC) can save a lot of effort.

Reason 3: The consistency of the theory ZFC is not more dubious than the consistency of the theory ZF.

Let us discuss these three points one by one. As for Reason 3: In 1938, Kurt Gödel proved that if ZF is a consistent theory, then so is ZFC. In Chapter 12, we shall give an outline of his proof.

Concerning Reason 2, let $[X]^2$ denote the set of unordered pairs of elements of X, and consider the following two statements:

(HSO) For every infinite set X, there exists an injection $f : X \times X \to X$;
(HSU) For every infinite set X, there exists an injection $f : [X]^2 \to X$.

Using the Axiom of Choice, it is trivial to show that (HSO) implies (HSU): Given X, let $f : X \times X \to X$ be an injection. For each unordered pair $\{x, y\}$ choose $g(\{x, y\})$ from the set $\{f(\langle x, y\rangle), f(\langle y, x\rangle)\}$.

The implication (HSO) $\to$ (HSU) is also a theorem of ZF: Given any set X, the function $h : X \to X \times X$ defined by $h(x) = \langle x, x\rangle$ is an injection. Thus, by the Cantor-Schröder-Bernstein Theorem, (HSO) implies that $X \approx X \times X$ for every infinite set X. By a result of Tarski[11], the latter implies the Axiom of Choice. Once we have proved (AC), we can deduce (HSU) as above.

This is of course a rather roundabout way of proving the implication (HSO) $\to$ (HSU). Here is a challenge for you:

EXERCISE 40(X): Find a simpler proof of the implication (HSO) $\to$ (HSU) in ZF.

By far the most important argument in favor of the Axiom of Choice is Reason 1. If ZF were the official framework of mathematics, then topologists would have to live without Tychonff's Theorem, algebraists would have to consider vector spaces without bases, and functional analysts could not prove the Hahn-Banach Theorem. To illustrate what can happen to set theory itself without the Axiom of Choice, consider the notion of a cardinal. In Chapter 11, we will show how one can (in ZFC) assign to each set X a set $|X|$ (its *cardinal number*) in such a way that $X \approx Y$ implies $|X| = |Y|$. An alternative construction of cardinal numbers can be carried out in ZF alone (see Definition 12.12). However, if cardinal numbers are supposed to conform to our basic intuitions about "numbers," then any two cardinals should be comparable. By Theorem 19, in the absence of AC, not all cardinals are pairwise comparable (by the discussion preceding Fact 18, even the word "comparable" is ambiguous in ZF). This phenomenon may lead to rather bizarre situations. For example, it is relatively consistent with ZF that there exists a set A such that $n < |A|$ for every natural number n, and yet there is no injection $f : \omega \to A$.

EXERCISE 41(PG): Show (in ZF) that the following are equivalent for every set A:

(a) There is no injection $f : \omega \to A$;
(b) Every injection $f : A \to A$ is a surjection.

A set A that satisfies conditions (a) and (b) of the above exercise is called *Dedekind-finite*. [12]

It is a matter of taste which one considers more counterintuitive: the existence of infinite, Dedekind-finite sets or the Banach-Tarski Paradox. But, definitely,

[11] See: A. Tarski, *Sur quelques théorèmes qui équivalent à l'axiome du choix*, Fund. Math. 5 (1924), 147–154.

[12] A discussion of various notions of finiteness in the absence of (AC) can be found in J. Truss, *Classes of Dedekind finite cardinals*, Fund. Math. 84 (1974) 187–208.

both situations *are* paradoxical. Could one perhaps add an axiom to ZF that would not imply the most paradoxical consequences of the Axiom of Choice, and yet would yield all its benefits? No. As Theorem 17 shows, if you want to have Tychonoff's Theorem, then you have to live with all consequences of (AC). Similarly, by Theorem 19, if you want all cardinal numbers to be comparable, then you must admit the Axiom of Choice. You cannot eat your cake and have it too. But you can eat some of it and leave the rest for later. Mathematicians have studied a number of weaker versions of the Axiom of Choice. Some of these are strong enough to prove that every Dedekind-finite set is finite, but do not imply the existence of nonmeasurable subsets of the real line. T. Jech's book *The Axiom of Choice*, North-Holland 1973, gives a thorough exposition of weaker versions of (AC). Let us discuss just two of them here.

The most natural way of weakening the Axiom of Choice is by restricting the cardinalities of the families involved. As an example of this procedure, let us formulate the *Countable Axiom of Choice.*

(AC)$_{\aleph_0}$ Every countable family A of nonempty, pairwise disjoint sets has a selector.

EXERCISE 42(G): (a) Return to the proof of Theorem 3.17 and convince yourself that the Axiom of Choice is tacitly used in it.[13]

(b) Convince yourself that Theorem 3.17 is provable in ZF + $(\mathrm{AC})_{\aleph_0}$.

EXERCISE 43(R): Prove in $(\mathrm{AC})_{\aleph_0}$ that every Dedekind-finite set is finite.

Your first impulse in attacking Exercise 43 was probably to go to the proof of Theorem 4.11 and convince yourself that it can be carried out in ZF + $(\mathrm{AC})_{\aleph_0}$. This is a reasonable thing to try, but it won't work. Our proof of Theorem 4.11 involves a recursive construction, and unfortunately, ZF + $(\mathrm{AC})_{\aleph_0}$ is not strong enough to prove the Principle of Recursive Constructions (Theorem 4.24). The latter theorem is equivalent to a weak form of the Axiom of Choice known as the *Principle of Dependent Choices.*

(DC) If R is a binary relation on a nonempty set X such that for every $x \in X$ there exists a $y \in X$ with $\langle x, y\rangle \in R$, then there exists a sequence $(x_n)_{n\in\omega}$ of elements of X such that $\langle x_n, x_{n+1}\rangle \in R$ for every $n \in \omega$.

EXERCISE 44(G): Prove that the Principle of Recursive Constructions and the Principle of Dependent Choices are equivalent in ZF.

EXERCISE 45(PG): (a) Dig out your solution of Exercise 2.10 and convince yourself that Theorem 2.2 is provable in ZF + (DC).

(b) Show that ZF + Theorem 2.2 $\vdash$ (DC).

EXERCISE 46(G): Show that ZF + (DC) $\vdash$ $(\mathrm{AC})_{\aleph_0}$.

The last three exercises demonstrate that the theory ZF + (DC) is a natural candidate for a foundations of mathematics. Another advantage of this framework is that it is apparently compatible with the most interesting alternative to the full Axiom of Choice proposed so far. This alternative is the topic of our next section.

---

[13] It can be shown that Theorem 3.17 is not a theorem of ZF.

9.3. The Axiom of Determinacy

J. Mycielski and H. Steinhaus proposed in 1962 an alternative to the Axiom of Choice which is called the *Axiom of Determinacy* and usually abbreviated (AD). It can be described in the form of infinite games.

Let $A \subseteq {}^{\omega}\omega$. Imagine a game Γ_A played by two persons, I and II, who take turns in picking natural numbers. Player I starts and chooses $a_0 \in \omega$, then Player II chooses $b_0 \in \omega$, then I chooses $a_1 \in \omega$, next II chooses $b_1 \in \omega$, and so on. If the resulting sequence $\langle a_0, b_0, a_1, b_1, \ldots \rangle$ is a member of A, then I wins, otherwise II wins. The sequence $\langle a_0, b_0, a_1, b_1, \ldots \rangle$ is called the *outcome* of the game. For $s \in {}^{\omega}\omega$, let $s^{\mathrm{I}} = (s_{2n})_{n<\omega}$ and $s^{\mathrm{II}} = (s_{2n+1})_{n<\omega}$. If s is the outcome of the game, then s^{I} is the sequence of numbers chosen by I, and s^{II} is the sequence of numbers chosen by II.

A *strategy* (for I or II) is a rule that tells the player which number to choose. The n-th move prescribed by a strategy depends on the moves made by the opponent up to this stage of the game. More specifically, a strategy for I is a function $F : \bigcup_{n\in\omega} {}^{n}\omega \to \omega$. Player I is said to *follow the strategy* F if she always chooses the number $F(b_0, \ldots , b_{n-1})$ for her n-th move a_n. Similarly, a strategy for II is a function $G : \bigcup_{n\in\omega\setminus\{0\}} {}^{n}\omega \to \omega$. Player II follows this strategy if he always chooses $b_n = G(a_0, \ldots , a_n)$.

A strategy is a *winning strategy,* if the player who follows it always wins, independently of what the opponent does. The game Γ_A is *determined* if one of the players has a winning strategy.

EXERCISE 47(G): Convince yourself that in any given game Γ_A *at most* one of the players can have a winning strategy.

EXERCISE 48(G): Let $A = \{s_i : i \in \omega\}$ be a countable subset of ${}^{\omega}\omega$. Let $G(a_0, \ldots , a_n) = s_n(2n) + 1$ for all $\langle a_0, \ldots , a_n\rangle$. Show that G is a winning strategy for player II in Γ_A.

EXERCISE 49(G): Let $B = \{s \in {}^{\omega}\omega : \forall n < \omega\, (s_{2n+1} + s_{2n+2} \text{ is even})\}$. Show that player I has a winning strategy in the game Γ_B.

The *Axiom of Determinacy* asserts the following:

(AD) For every $A \subseteq {}^{\omega}\omega$ the game Γ_A is determined.

Let $n < \omega$ and $t \in {}^{n}\omega$. We define $U_t = \{s \in {}^{\omega}\omega : t \subseteq s\}$. The family $\mathcal{B} = \{U_t : t \in \bigcup {}^{n}\omega\}$ is a basis for a topology τ on ${}^{\omega}\omega$. This means that a set X is in τ if and only if X is a union of elements of $\mathcal{B}$.

EXERCISE 50(PG): Show that $\langle {}^{\omega}\omega, \tau\rangle$ is a second countable completely metrizable space. *Hint:* For $r, s \in {}^{\omega}\omega$ such that $r \neq s$, let $\varrho(r, s) = 1/\min\{n : r(n) \neq s(n)\}$. Moreover, let $\varrho(r, r) = 0$. Show that ϱ is a complete metric on ${}^{\omega}\omega$ that induces the topology τ.

If player I follows a strategy F, then the outcome depends only on F and the sequence $b = (b_n)_{n<\omega}$ of moves played by II. We denote the outcome of such a game by $F * b$. If player II follows a strategy G, then the outcome depends only on G and the sequence $a = (a_n)_{n<\omega}$ of player I's moves. We denote the outcome of such a game by $a * G$. If player I follows a strategy F and player II follows a strategy G, then we denote the outcome of the resulting game by $F * G$.

Let F be a strategy for player I. By $\widehat{F}$ we denote the set $\{F * b : b \in {}^{\omega}\omega\}$ of all outcomes compatible with F. Similarly, if G is a strategy for player II, then we let $\widehat{G} = \{a * G : a \in {}^{\omega}\omega\}$.

EXERCISE 51(G): Show that if F is a strategy for player I and G is a strategy for player II, then $\widehat{F}$ and $\widehat{G}$ are uncountable closed subsets of $\langle X, \tau \rangle$.

Theorem 35 (ZF + (AD)): *There exists no Bernstein set in* $\langle {}^{\omega}\omega, \tau \rangle$.

Proof: Let $A \subset {}^{\omega}\omega$. By (AD), one of the players has a winning strategy in Γ_A. If F is a winning strategy for player I, then $\widehat{F} \subseteq A$. If G is a winning strategy for player II, then $\widehat{G} \cap A = \emptyset$. In either case, by Exercise 51, A is not a Bernstein set. □

Corollary 36 (ZF): (AD) $\rightarrow \neg$(AC).

Proof: By Theorem 11.4 and Exercise 50, (AC) implies the existence of a Bernstein set in $\langle {}^{\omega}\omega, \tau \rangle$. □

Although (AD) is incompatible with the full Axiom of Choice, it implies a weak version of the latter.

Lemma 37 (ZF + (AD)): *Let* $X = \{X_i : i < \omega\}$ *be a countable family of nonempty subsets of* ${}^{\omega}\omega$. *Then* $\prod_{i<\omega} X_i \neq \emptyset$.

Proof: Let A be the set of all $s \subseteq {}^{\omega}\omega$ with $s^{\mathrm{II}} \in X_{s(0)}$. Let $B = {}^{\omega}\omega \setminus A$, and consider the game G_B. Player I cannot have a winning strategy in this game: Suppose F is a strategy for I, and suppose F prescribes a_0 as I's first move. If $b \in X_{a_0}$, then $F * b \in A$, and player II wins the game. Since X_{a_0} is not empty, it is possible for player II to defeat F.

Since player I does not have a winning strategy in Γ_A, it follows from (AD) that player II has a winning strategy G. We fix such G and define a choice function $f \in \prod_{i<\omega} X_i$ by $f(i) = (\langle i, 0, 0, \ldots \rangle * G)^{\mathrm{II}}$.

EXERCISE 52(G): Convince yourself that the function f defined in the last line of the above proof is an element of $\prod_{i<\omega} X_i$. □

The definition of Γ_A can be generalized in a straightforward manner to a situation where X is an arbitrary nonempty set and $A \subseteq {}^{\omega}X$. The Axiom of Determinacy implies that for each countable set X and each $A \subseteq {}^{\omega}X$, the game Γ_A is determined.

Theorem 38 (ZF + (AD)): *There is no free ultrafilter on* ω.

Proof: Assume toward a contradiction that $U \subseteq \mathcal{P}(\omega)$ is a free ultrafilter on ω. We shall describe a game Γ_A in which neither of the two players has a winning strategy.

Let $X = [\omega]^{<\aleph_0}$. Both players of Γ_A choose elements from X. It is the aim of player I to choose a_n's in such a way that $\bigcup_{n<\omega} a_n \in U$. If he can do so he wins; otherwise player II wins. The rules of the game demand that all sets a_n, b_n are pairwise disjoint. In other words, the player who first chooses a set a_n or b_n that is not disjoint from all sets a_k, b_k chosen so far has lost the game.

EXERCISE 53(PG): Define a set $A \subseteq {}^{\omega}X$ so that the game described above is Γ_A.

Now suppose that player I has a winning strategy. We define a strategy for player II by

$$G(a_0, \cdots, a_n) = \begin{cases} \emptyset & \text{if } n = 0; \\ F(a_1, \ldots, a_n) \setminus a_0 & \text{otherwise.} \end{cases}$$

This means, player II starts with the empty set. Ignoring the first move of player I, he emulates the strategy of player I (but he deletes a_0 to ensure that the sets he chooses are disjoint from all sets chosen so far).

Consider the outcome $F * G = \langle a_0, b_0, a_1, b_1, \ldots \rangle$. Since F was supposed to be a winning strategy for player I, $\bigcup_{n\in\omega} a_n \in U$. Moreover, $F * \langle a_1, a_2, \ldots \rangle = \langle a_0, a_1, F(a_1), a_2, F(a_1, a_2), \ldots \rangle = \langle a_0, a_1, b_1, a_2, b_2, \ldots \rangle$ is also a winning outcome for player I, and hence $a_0 \cup \bigcup_{n\in\omega\setminus\{\emptyset\}} b_n \in U$. But $(a_0 \cup \bigcup_{n\in\omega\setminus\{\emptyset\}} b_n) \cap (\bigcup_{n\in\omega} a_n) = a_0$, which leads to a contradiction, since U was supposed to be free.

Now suppose that player II has a winning strategy G. We define a strategy G^* by

$$G^*(t_0, \cdots, t_n) = \begin{cases} G(t_0, \cdots, t_n) \setminus n, & \text{if } n \in \bigcup_{k\le n} t_k \cup \bigcup_{k<n} G(t_0, \cdots, t_k); \\ (G(t_0, \cdots, t_n) \setminus n) \cup \{n\} & \text{otherwise.} \end{cases}$$

Then G^* is also a winning strategy for player II. To see this, consider the outcome $a * G^* = \langle a_0, b_0^*, a_1, b_1^*, \ldots \rangle$ of a game in which player I respects the rule of never choosing any number already previously chosen by either of the players. Note that there is an outcome $\langle a_0, b_0, a_1, b_1, \ldots \rangle = a * G$, where player II follows strategy G, player I plays exactly as before, and neither player violates the rules. Moreover, since $\bigcup_{n\in\omega} a_n \cup \bigcup_{n\in\omega} b_n^* = \omega$ and $\bigcup_{n\in\omega} a_n \notin U$, it must be the case that $\bigcup_{n\in\omega} b_n^* \in U$.

Now define a strategy F for player I by

$$F(b_0, \cdots, b_{n-1}) = G^*(\emptyset, b_0, \cdots, b_{n-1}).$$

Let $\langle a_0, b_0^*, a_1, b_1^*, \ldots \rangle = F * G$. Then $\bigcup_{n\in\omega} b_n^* \in U$. But the sequence $\langle \emptyset, a_0, b_0^*, a_1, b_1^*, \ldots \rangle$ is also an outcome of a game in which player II follows G^*, and hence $\bigcup_{n\in\omega} a_n \in U$. Since $(\bigcup_{n\in\omega} a_n) \cap (\bigcup_{n\in\omega} b_n) = \emptyset$, we get again a contradiction. □

One might expect that for reasonably "nice" subsets A of ${}^\omega\omega$ it can be shown in ZFC that the game Γ_A is determined. In 1975, D. A. Martin showed in ZFC that the game Γ_A is determined whenever A is a Borel subset[14] of $\langle {}^\omega\omega, \tau \rangle$. Since open sets are Borel, the following theorem, which was obtained in 1953 by D. Gale and F. M. Stewart, is a special case of Martin's theorem.

Theorem 39 (ZFC): *Let A be an open subset of ${}^\omega\omega$. Then the game Γ_A is determined.*

Proof: We need to introduce a few new symbols. Let $s \in {}^\omega\omega$, $n \in \omega$, $A \subseteq {}^\omega\omega$, $t \in {}^n\omega$. We define $s \dot{-} n = \langle s(n), s(n+1), \ldots \rangle$ and $A^t = \{s \dot{-} n : s \in A, t \subseteq s\}$. Now let A be an open subset of ${}^\omega\omega$. Assume that Player I does not have a winning strategy in Γ_A. We construct a winning strategy for II as follows: Suppose player I starts with a_0. Since I does not have a winning strategy, there exists a b_0 such that I does not have a winning strategy in the game $\Gamma_{A^{\langle a_0, b_0 \rangle}}$. Player II chooses

[14] Borel subsets of topological spaces will be discussed in Chapter 13 of Volume II.

such a b_0. Now suppose I continues with a_1. Then II chooses b_1 such that I does not have a winning strategy $\Gamma_{A\langle a_0,b_0,a_1,b_1\rangle}$, and so on.

We show that playing this way, II wins the game. Let $s = \langle a_0, b_0, a_1, b_1, \ldots\rangle$ be the outcome of a game in which II follows the above strategy. We have to show that $s \notin A$. If $s \in A$, there exists $t = \langle a_0, b_0, \ldots, a_n, b_n\rangle \subset s$ such that $U_t \subseteq A$. But then player II has already lost the game when the configuration $\langle a_0, b_0, \ldots, a_n, b_n\rangle$ is reached, which contradicts the choice of $b_0, \ldots, b_n$. □

EXERCISE 54(G): Show that if A is a closed subset of ${}^{\omega}\omega$, then the game Γ_A is determined.

Not only is Γ_A determined whenever A is "nice," it also works the other way around: If Γ_A is determined for every $A \subset {}^{\omega}\omega$, then all subsets of ${}^{\omega}\omega$ have rather desirable properties: For example, (AD) implies that all subsets of the real line are Lebesgue measurable and have the Baire Property.[15] Thus it makes sense to consider the possibility of developing mathematics in the theory ZF + (AD), or even better, in ZF + (DC) + (AD). But are these theories consistent? As it turned out, the theory ZF + (DC) + (AD) is equiconsistent with the existence of certain large cardinals. In Chapter 12 we will explain what this means. For now, let us just say that the consistency of ZF + (AD) is somewhat more doubtful than the consistency of ZF itself, but most set theorists would take equiconsistency with a large cardinal as strong evidence that the theory ZF + (DC) + (AD) is indeed consistent.

9.4. The Banach-Tarski Paradox

In this section we give a proof of the Banach-Tarski Paradox. It is assumed that the reader has some knowledge of group theory.

In his book *Grundzüge der Mengenlehre,* Leipzig, 1914, F. Hausdorff constructed three disjoint subsets A, B, and C of the sphere S such that $S \setminus (A \cup B \cup C)$ is countable, $A \cong B \cong C$, and also $A \cong B \cup C$ (where $X \cong Y$ means that X and Y are congruent, i.e., there exists an isometry ϕ of the sphere such that $\phi[X] = Y$).

We start with a lemma on rotations. Let $G = G_0 * G_1$ be the free product of two groups $G_0 = \{e, \varphi\}$ und $G_1 = \{e, \psi, \psi^2\}$. Note that these assumptions imply that $\varphi^2 = e$ and $\psi^3 = e$. Thus, the elements of G can be represented by finite sequences in which φ alternates with ψ and ψ^{-1} $(= \psi^2)$. For instance, $\varphi \circ \psi^{-1} \circ \varphi \circ \psi \circ \varphi$ is such a sequence. We will refer to these sequences as *irreducible representations* of elements of G.

Let d_0 and d_1 be axes of rotation going through the center of the sphere S. Let φ_0 be a rotation by 180° about d_0 and ψ_0 a rotation by 120° about d_1. We consider the group which is generated by φ_0 and ψ_0. This group depends on the angle δ between d_0 and d_1. There is a canonical homomorphism π from G onto the group generated by φ_0 and ψ_0. For each $\alpha \in G$ there are only finitely many δ such that $\pi(\alpha)$ is the identity. Thus we can find δ such that π is an isomorphism. We pick such a δ and identify from now on G and $\pi[G]$.

Lemma 40: *There is a partition $\{A, B, C\}$ of G such that $A \circ \varphi = B \cup C$, $A \circ \psi = B$, and $A \circ \psi^{-1} = C$.*

---

[15]The Baire Property will be defined in Chapter 13 of Volume II.

Proof: We define A, B, and C by induction on the length of the elements of G. Let

$$e \in A, \qquad \varphi, \psi \in B, \qquad \psi^{-1} \in C.$$

Assume that α is of length ≥ 1 and ends with ψ or ψ^{-1}. Then:

(1) if $\alpha \in A$, then $\alpha \circ \varphi \in B$;

(2) if $\alpha \in B \cup C$, then $\alpha \circ \varphi \in A$.

If α ends with φ, then:

(3) if $\alpha \in A$, then $\alpha \circ \psi \in B$ and $\alpha \circ \psi^{-1} \in C$;

(4) if $\alpha \in B$, then $\alpha \circ \psi \in C$ and $\alpha \circ \psi^{-1} \in A$;

(5) if $\alpha \in C$, then $\alpha \circ \psi \in A$ and $\alpha \circ \psi^{-1} \in B$.

We show that $A \circ \varphi = B \cup C$ and leave the other equations as an exercise.

First we show that $A \circ \varphi \subseteq B \cup C$.

Assume that $\alpha \in A$, where α is an irreducible representation of an element of G. If α ends with ψ or ψ^{-1}, then it follows from (1) that $\alpha \circ \varphi \in B$.

Assume that α ends with φ, that is, $\alpha = \beta \circ \varphi$. Then by (2), $\beta \in B \cup C$, and since $\alpha \circ \varphi = \beta \circ \varphi \circ \varphi = \beta$, we have $\alpha \circ \varphi \in B \cup C$. Thus we have shown that $A \circ \varphi \subseteq B \cup C$.

Now we show that $B \cup C \subseteq A \circ \varphi$. Let α be an irreducible representation of an element of $B \cup C$.

If α ends with ψ or ψ^{-1}, then $\alpha \circ \varphi \in A$ by (2), and thus $\alpha = \alpha \circ \varphi \circ \varphi \in A \circ \varphi$.

Now suppose $\alpha = \beta \circ \varphi$. Then β must be the empty string or end with ψ or ψ^{-1}. In the former case, $\alpha \in A \circ \varphi$ since $e \in A$. In the latter case, by clause (2) again, β must be an element of A, and thus $\alpha \in A \circ \varphi$, as desired. Thus we have shown that also $B \cup C \subseteq A \circ \varphi$, and hence $B \cup C = A \circ \varphi$. □

EXERCISE 55(G): Prove the remaining equations in Lemma 40.

Theorem 41: *There is a partition* $\{X, Y, Z, Q\}$ *of* S *with* $|Q| = \omega$ *such that*
(i) $X \cong Y \cong Z$;
(ii) $X \cong Y \cup Z$.

Proof: Let φ and ψ as in Lemma 40. Each rotation of the sphere that is not the identity has exactly two fixed points. Thus the set Q of all points of S which are fixed points for some $\alpha \in G \setminus \{e\}$ is countable.

For each $x \in S \setminus Q$ let

$$P_x = \{\alpha(x) : \alpha \in G\}$$

be the orbit of x. Then $S \setminus Q$ is the disjoint union of the orbits. By the Axiom of Choice, there exists a set M that intersects each of these orbits in exactly one point. Let

$$X = \{\alpha(a) : \alpha \in A, a \in M\};$$
$$Y = \{\alpha(a) : \alpha \in B, a \in M\};$$
$$Z = \{\alpha(a) : \alpha \in C, a \in M\}.$$

Then by Lemma 40, $\varphi[X] = Y \cup Z$, $\psi[X] = Y$ and $\psi^{-1}[X] = Z$. It follows that $X \cong Y \cong Z$ and $X \cong Y \cup Z$. □

Now we show how Hausdorff's result can be used to prove the Banach-Tarski Paradox.

Define a relation $\approx$ on $\mathcal{P}(\mathbb{R}^3)$ by: $X \approx Y$ if there are $n < \omega$ and partitions $(X_i)_{i<n}$ of X and $(Y_i)_{i<n}$ of Y such that $X_i \cong Y_i$ for all $i < n$.

Lemma 42: (i) $\approx$ *is an equivalence relation;*

(ii) *if* $(X_i)_{i<n}$ *is a partition of* X *and* $(Y_i)_{i<n}$ *is a partition of* Y *with* $X_i \approx Y_i$ *for all* $i < n$*, then* $X \approx Y$*;*

(iii) *if* $X' \subseteq Y \subseteq X$ *and* $X' \approx X$ *then also* $Y \approx X$.

Proof: (i) and (ii) are easily seen.

For the proof of (iii), let $(X_i')_{i<n}$ and $(X_i)_{i<n}$ be partitions of X' and X such that $X_i' \cong X_i$ for all $i < n$. For each $i < n$ let $f_i : X_i \to X_i'$ be an isometry with $f_i[X_i] = X_i'$. Let $f = \bigcup_{i<n} f_i$. We set

$$X^0 = X,\ X^1 = f[X^0],\ \dots,\ X^{n+1} = f[X^n],\ \dots,$$
$$Y^0 = Y,\ Y^1 = f[Y^0],\ \dots,\ Y^{n+1} = f[Y^n],\ \dots .$$

Consider the set

$$Z = \bigcup_{n<\omega} (X^n \setminus Y^n).$$

One can show by induction over n that $f[Z] \subseteq Z$. Now $X = Z \cup (X \setminus Z)$, $Y = f[Z] \cup (X \setminus Z)$, and $Z \approx f[Z]$ implies $Y \approx X$. □

Theorem 43 (Banach and Tarski): *The unit ball* U *can be partitioned into two sets* X *and* Y *with* $X \approx U$ *and* $Y \approx U$.

Proof: The unit sphere S is the boundary of U. Let c denote the center of U. For $D \subseteq S$ let $\widehat{D}$ denote the set of all points $a \in U \setminus \{c\}$ whose projection from c on S is in D.

Let $\{A, B, C, Q\}$ be a partition of S as in Theorem 41. Then $\{\widehat{A}, \widehat{B}, \widehat{C}, \widehat{Q}, \{c\}\}$ is a partition of U. By Theorem 41 we have $\widehat{A} \approx \widehat{B} \approx \widehat{C} \approx \widehat{B} \cup \widehat{C}$. Thus, by Lemma 42(ii), $\widehat{B} \cup \widehat{C} \approx \widehat{A} \cup (\widehat{B} \cup \widehat{C}) \approx \widehat{A} \cup \widehat{B} \cup \widehat{C}$, and transitivity of $\approx$ implies that $\widehat{C} \approx \widehat{A} \cup \widehat{B} \cup \widehat{C}$. Similarly, $\widehat{A} \approx \widehat{A} \cup \widehat{B} \cup \widehat{C}$ and $\widehat{B} \approx \widehat{A} \cup \widehat{B} \cup \widehat{C}$.
Hence,

$$\widehat{A} \cup \widehat{Q} \cup \{c\} \approx U. \tag{1}$$

Let α be a rotation that is not in G such that Q and $\alpha[Q]$ are disjoint. Then $\alpha[Q] \subseteq A \cup B \cup C$, and $\widehat{C} \approx \widehat{A} \cup \widehat{B} \cup \widehat{C}$ implies that there is a $T \subseteq C$ with $Q \approx T$. Let $p \in C \setminus T$. Then

$$\widehat{A} \cup \widehat{Q} \cup \{c\} \approx \widehat{B} \cup \widehat{T} \cup \{p\}. \tag{2}$$

Furthermore, $\widehat{B} \cup \widehat{T} \cup \{p\} \subseteq \widehat{B} \cup \widehat{C} \subseteq U$. By (1) and (2), $\widehat{B} \cup \widehat{C} \approx U$. Let

$$X = \widehat{A} \cup \widehat{Q} \cup \{c\}, \qquad Y = \widehat{B} \cup \widehat{C}.$$

Then $\{X, Y\}$ is the partition of U we are looking for. □

EXERCISE 56(PG): Show that if A is a bounded subset of $\mathbb{R}^3$ with nonempty interior, then $A \approx U$.

CHAPTER 10

The Ordinals

10.1. The Class ON

It is important to know that some results of this section are provable without the help of the Axiom of Choice. Therefore, we shall work in ZF throughout this section. In order to facilitate reference, some theorems will be explicitly marked as ZF-results.

We shall freely go back and forth between considering p.o.'s and their corresponding strict p.o.'s. In particular, the term "order isomorphism" that was originally defined only for p.o.'s will also be used for strict partial orders. Despite the fact that a p.o. $\langle X, \leq \rangle$ and its corresponding strict partial order $\langle X, < \rangle$ are non-isomorphic models of the language $L_{\leq}$, we define $ot(\langle X, \leq \rangle)$ and $ot(\langle X, < \rangle)$ as the same object. This approach may look inelegant and may also be slightly confusing to the beginner, but it truthfully reflects mathematical practice.

Definition 1: A set α is called an *ordinal* if α is transitive and strictly wellordered by $\in$.

EXERCISE 1(G): Convince yourself that α is an ordinal iff α is a transitive set and $\forall \beta, \gamma \in \alpha \, (\beta \in \gamma \vee \beta = \gamma \vee \gamma \in \beta)$.

Let $\langle X, \prec \rangle$ be a strict w.o. Then $\prec$ is a strictly wellfounded relation on X. By Exercise 4.55, $\prec$ is also extensional. By Theorem 4.39 and Exercise 4.57, there exist exactly one transitive set α, the Mostowski collapse of $\langle X, \prec \rangle$, and a unique bijection $F : X \to \alpha$, called the *collapsing function,* such that

$$\text{(M)} \qquad \forall z, z' \in Z \, (z \prec z' \leftrightarrow F(z) \in F(z')).$$

Theorem 2: *A set α is an ordinal iff α is the Mostowski collapse of a strict w.o. $\langle X, \prec \rangle$.*

EXERCISE 2(PG): (a) Show that if α is the Mostowski collapse of a strict w.o. $\langle X, \prec \rangle$, then α is an ordinal, and the collapsing function is an order isomorphism between $\langle X, \prec \rangle$ and $\langle \alpha, \in \rangle$.

(b) Show that if α is an ordinal, then α is the Mostowski collapse of $\langle \alpha, \in \rangle$, and the collapsing function is the identity.

(c) Conclude that Theorem 2 holds. □

EXERCISE 3(PG): Show that if $\langle X, \prec_X \rangle$ and $\langle Y, \prec_Y \rangle$ are isomorphic strict p.o.'s, then they have the same Mostowski collapse.

How convenient! Let

$$\mathbf{WOT} = \{\langle x, y\rangle : x \text{ is a strict w.o. and } y \text{ is the Mostowski collapse of } x\}.$$

Then **WOT** is a functional class that satisfies the following conditions:

(i) If $\langle x, y\rangle \in \mathbf{WOT}$, then x and $\langle y, \in\rangle$ are isomorphic strict w.o.'s.

(ii) The domain of **WOT** is the class of all strict w.o.'s.

(iii) For all strict w.o.'s x, y, and ordinals z, if $\langle x, z\rangle \in \mathbf{WOT}$ and $x \cong y$, then $\langle y, z\rangle \in \mathbf{WOT}$.

Compare these conditions with conditions (i)–(iii) of Exercise 8.16. The class **WOT** behaves exactly as the class **OT**, except that the domain of **WOT** is the class of all strict w.o.'s (rather than the class of all p.o.'s), and the class **WOT** can be defined in ZF, without reference to a wellorder of the whole universe.

In Section 8 we have seen that the naive concept of an order type leads to conceptual difficulties: If the order type of $\langle X, \preceq_X\rangle$ is the collection of all $\langle Y, \preceq_Y\rangle$ which are order-isomorphic to $\langle X, \preceq_X\rangle$, then order types are proper classes and such notions as addition and multiplication of order types cannot be defined within our axiomatic framework. Therefore, let us from now on use the following definition of the order type of a wellorder.

Definition 3: (a) Let $\langle X, \prec\rangle$ be a strict w.o. We define the *order type* of $\langle X, \prec\rangle$, written $ot(\langle X, \prec\rangle)$, as the unique ordinal α such that $\langle X, \prec\rangle \cong \langle \alpha, \in\rangle$.

(b) Let $\langle X, \preceq\rangle$ be a w.o., and let $\prec = \{\langle x, y\rangle \in X \times X : x \preceq y \wedge x \neq y\}$ be the strict wellorder relation corresponding to $\preceq$. We define the *order type* $ot(\langle X, \preceq\rangle)$ by setting $ot(\langle X, \preceq\rangle) = ot(\langle X, \prec\rangle)$.

Note that $ot(\langle X, \preceq\rangle) = \alpha$ if and only if $\langle\langle X, \prec\rangle, \alpha\rangle \in \mathbf{WOT}$. Ordinals are usually denoted by lower case Greek letters. The class of all ordinals is denoted by **ON**.

The next lemma shows that every element of an ordinal is an ordinal.

Lemma 4: $\forall \alpha \in \mathbf{ON}\, (\alpha \subseteq \mathbf{ON})$.

Proof: Assume $\alpha \in \mathbf{ON}$ and $\beta \in \alpha$. Since α is transitive, $\beta \subset \alpha$. By Exercise 2.15, the restriction of a strict wellorder relation to any subset of its domain is a strict wellorder relation. Thus, β is strictly wellordered by $\in$.
To show that β is a transitive set, assume $\gamma \in \beta$ and $\delta \in \gamma$. We need to show that $\delta \in \beta$. By transitivity of α, both γ and δ are elements of α. Moreover, since α is strictly wellordered by $\in$, we have

(i) $\delta \in \beta$; or

(ii) $\beta \in \delta$; or

(iii) $\beta = \delta$.

EXERCISE 4(G): Show that the last two possibilities are ruled out by the Foundation Axiom.

It follows that (i) holds; and Lemma 4 is proved. □

Corollary 5: *Every initial segment of an ordinal is an ordinal.*

Proof: Let $\alpha \in \mathbf{ON}$, and let $I \subseteq \alpha$ be an initial segment of the strict w.o. $\langle \alpha, \in \rangle$. If $I = \alpha$, then there is nothing to prove. If I is a proper initial segment of $\langle \alpha, \in \rangle$, then $I = I_{\in}(\zeta)$ for some $\zeta \in \alpha$ (see 4.20 for the definition of $I_{\in}(\zeta)$ and Exercise 4.42(d) for the analogous result for w.o.'s). But since the strict wellorder relation is $\in$, we have $I_{\in}(\zeta) = \{\beta \in \alpha : \beta \in \zeta\}$. By transitivity of α, the latter set is equal to $\{\beta : \beta \in \zeta\} = \zeta$. Since $\zeta \in \alpha$, Lemma 4 implies that $\zeta \in \mathbf{ON}$. $\square$

Lemma 6: $\forall \alpha, \beta \in \mathbf{ON}\, (\alpha \in \beta \vee \beta \in \alpha \vee \alpha = \beta)$.

Proof: Let $\alpha, \beta \in \mathbf{ON}$. By Theorem 4.30, either $\langle \alpha, \in \rangle$ is order-isomorphic to an initial segment of $\langle \beta, \in \rangle$, or $\langle \beta, \in \rangle$ is order-isomorphic to an initial segment of $\langle \alpha, \in \rangle$. Without loss of generality, assume the former, and let $F : \alpha \rightarrow \beta$ be a one-to-one function that maps α onto an initial segment I of β and such that

$$(*) \qquad \xi \in \eta \text{ iff } F(\xi) \in F(\eta) \text{ for all } \xi, \eta \in \alpha.$$

By Corollary 5, $rng(F)$ is an ordinal. In particular, the range of F is a transitive set. It is not hard to see that condition $(*)$ is exactly the same as condition (M) for the collapsing function of $\langle \alpha, \in \rangle$. Thus, $rng(F)$ is the Mostowski collapse of $\langle \alpha, \in \rangle$. By Exercise 2(b), $rng(F) = \alpha$. If F is a surjection, then $\alpha = rng(F) = \beta$. If not, then by Exercise 4.42(d), $\alpha = rng(F) = I_{\in}(\zeta) = \zeta$ for some $\zeta \in \beta$. In other words, in the latter case $\alpha \in \beta$. $\square$

EXERCISE 5(PG): Prove Lemma 6 without reference to the Mostowski collapse.

Corollary 7: *The relational class* $\in \cap (\mathbf{ON} \times \mathbf{ON})$ *strictly wellorders* $\mathbf{ON}$.

EXERCISE 6(PG): Recall from Chapter 8 what is meant by the phrase "a relational class $\mathbf{W}$ wellorders a class $\mathbf{X}$." Derive Corollary 7 from Lemma 6. $\square$

If α and β are ordinals, we shall often write $\alpha < \beta$ instead of $\alpha \in \beta$. This terminology is justified by Corollary 7. Thus, $\alpha \leq \beta$ if and only if α is an initial segment of β. (Compare this with Definition 4.33.)

Claim 8: *Let* $\alpha, \beta \in \mathbf{ON}$. *Then* $\alpha \leq \beta$ *iff* $\alpha \subseteq \beta$.

Proof: Let $\alpha, \beta \in \mathbf{ON}$. In terms of the notation $<$, Lemma 4 states that $\alpha = \{\gamma \in \mathbf{ON} : \gamma < \alpha\}$ and $\beta = \{\gamma \in \mathbf{ON} : \gamma < \beta\}$. From this representation it is evident that $\alpha \leq \beta$ implies $\alpha \subseteq \beta$. For the other implication, assume $\alpha \subseteq \beta$. If $\alpha = \beta$, there is nothing to prove. Thus, assume $\alpha \neq \beta$, and let $\gamma \in \beta \backslash \alpha$. Since $\gamma \notin \alpha$, we have either $\alpha = \gamma$ or $\alpha \in \gamma \in \beta$. Since β is a transitive set, in both cases we can infer that $\alpha \leq \beta$. $\square$

Now suppose $\langle X, \preceq \rangle$ is a w.o., and assume $y \notin X$. By Claim 9.15, if $Y = X \cup \{y\}$ and $\preceq_Y = \preceq \cup \{\langle x, y \rangle : x \in Y\}$, then $\langle Y, \preceq_Y \rangle$ is also a w.o.

EXERCISE 7(PG): Show that in the above notation, if $\alpha = ot(\langle X, \preceq\rangle)$, then $S(\alpha) = ot(\langle Y, \preceq_Y\rangle)$.

Exercise 7 shows that for every ordinal α, the successor $S(\alpha)$ is also an ordinal. Moreover, since $S(\alpha) = \alpha \cup \{\alpha\}$, we have $\alpha \in S(\alpha)$, and if $\beta \in S(\alpha)$, then $\beta \in \alpha$ or $\beta = \alpha$. Therefore, $S(\alpha)$ is the smallest ordinal bigger than α, i.e., $S(\alpha)$ is the *immediate successor* of α in **ON**.

Now let us see how the first few elements of **ON** look. The empty set is an ordinal. To see this, you may either verify that $\emptyset$ vacuously satisfies Definition 1, or convince yourself that $\emptyset$ is the Mostowski collapse of the empty strict w.o. $\langle\emptyset, \emptyset\rangle$. Moreover, since $\emptyset$ does not contain any elements, it must be the smallest ordinal. The next ordinal is $S(\emptyset)$, then come $S(S(\emptyset))$, $S(S(S(\emptyset)))$, etc. We encountered these sets already earlier in this text, didn't we? In Chapter 4, we denoted the n-th successor of $\emptyset$ by $\tilde{n}$ and identified it with the natural number n. Now it turns out that each $\tilde{n}$ is an ordinal. Moreover, the relation $<$ restricted to the ordinals $\tilde{n}$ is the natural strict order of the natural numbers, and $\tilde{n} = ot(\langle\{0, 1, \dots, n-1\}, <\rangle)$. Thus, the natural numbers as described in Chapter 4 are the finite ordinals.

EXERCISE 8(PG): Show that an ordinal α is a finite set iff it is a natural number.

The smallest infinite ordinal is the set ω of all natural numbers. We know already that ω is an infinite set. By Exercise 4.4(c), ω is transitive. Moreover, whenever n, m are natural numbers,[1] then either $n \in m$, or $m \in n$, or $n = m$. By Exercise 2(a), ω is an ordinal. (Alternatively, by the solution to Exercise 4.62(a), ω is the Mostowski collapse of $\langle\omega, <\rangle$. By Theorem 2, ω is an ordinal.)

To see that ω is the *smallest* infinite ordinal, consider an arbitrary infinite ordinal α. Since all elements of ω are finite sets, it is not the case that $\alpha \in \omega$. By Lemma 6, either $\omega = \alpha$ or $\omega \in \alpha$; i.e., $\omega \leq \alpha$.

EXERCISE 9(G): Find some other infinite ordinals.

The fact that ω is the smallest infinite ordinal allows us to characterize the set of natural numbers without self-reference.

Lemma 9 (ZF): *ω is the least ordinal α such that*
(i) *$\alpha \neq 0$, and*
(ii) *$S(\beta) \in \alpha$ for all $\beta \in \alpha$.*

EXERCISE 10(PG): Go back to Corollary 7.4 and convince yourself that the parameter ω can be dropped from the definition of $S^{(n)}(x)$.

The construction of the ordinal ω from the set of all finite ordinals can be generalized as follows.

Theorem 10: *Let A be a set of ordinals. Then:*
(a) *$\bigcup A \in$ **ON**;*

[1] As in Chapter 4, we shall write just n instead of $\tilde{n}$.

(b) *If* $\forall \alpha \in A \exists \beta \in A\,(\alpha < \beta)$, *then* $\bigcup A$ *is the smallest ordinal that exceeds all ordinals in* A.

EXERCISE 11(G): Prove point (a) of the above theorem.

Proof of Theorem 10(b): Let A be as in the assumptions, and let $\delta = \bigcup A$. Then $\delta = \{\alpha : \exists \beta \in A\,(\alpha \in \beta)\}$, and hence the assumption of point (b) implies that $\alpha \in \delta$ for all $\alpha \in A$. On the other hand, if an ordinal α is in δ, then α is in β for some $\beta \in A$. Thus, no ordinal smaller than δ contains all elements of A.$\square$

If A is a set of ordinals as in the assumptions of Theorem 10(b), then we shall often write sup A instead of $\bigcup A$. If $A = \{\alpha_\beta : \beta < \gamma\}$ and $\alpha_\beta \leq \alpha_{\beta'}$ for $\beta < \beta' < \gamma$, then we frequently write $\lim_{\beta \to \gamma} \alpha_\beta$ instead of $\sup\{\alpha_\beta : \beta < \gamma\}$.

EXERCISE 12(G): (a) Show that if A is a set of ordinals that contains a largest element α, then $\bigcup A \leq \alpha$.

(b) Give examples of A and α as in point (a) such that

(b1) $\bigcup A = \alpha$;

(b2) $\bigcup A < \alpha$.

Corollary 11: ON *is a proper class.*

Proof: First note that there is no largest ordinal: If α is any ordinal, then $S(\alpha)$ is a larger ordinal. Now suppose toward a contradiction that **ON** is a set. Since **ON** has no largest element, the assumptions of Theorem 10(b) are satisfied. But then $\bigcup$ **ON** is an ordinal that is larger than every ordinal, which is impossible.$\square$

Note that the crucial point in deriving the above contradiction was the assumption that **ON** is a set: by the Union Axiom, $\bigcup A$ is a set for every set A. If **ON** is a proper class, then none of the axioms of ZFC entitles us to form a set $\bigcup$ **ON**, and we get no contradiction.

Definition 12: A class $\mathbf{X} \subseteq \mathbf{ON}$ is *cofinal in* **ON** if $\forall \alpha \in \mathbf{ON} \exists \beta \in \mathbf{X}\,(\alpha \leq \beta)$. An ordinal α is called a *successor ordinal* if there exists an ordinal β such that $\alpha = S(\beta)$. An ordinal which is not a successor ordinal is called a *limit ordinal.* We denote the class of all successor ordinals by **SUCC**, and the class of all limit ordinals by **LIM**.

The positive natural numbers $1, 2, 3, \ldots$ are successor ordinals, while 0 and ω are examples of limit ordinals. It is easy to see that the class of successor ordinals is cofinal in **ON**: If $\alpha \in \mathbf{ON}$, then $\alpha < S(\alpha) \in \mathbf{SUCC}$.

EXERCISE 13(G): (a) Show that if a set A of ordinals satisfies the assumptions of Theorem 10(b), then $\bigcup A$ is a limit ordinal.

(b) Show that the class **LIM** is cofinal in **ON**.

Theorem 13: *Every cofinal subclass of* **ON** *is proper. In particular, both* **SUCC** *and* **LIM** *are proper classes.*

EXERCISE 14(PG): Prove Theorem 13.

Let us illustrate how the use of ordinals can simplify the study of w.o.'s and related concepts. In Chapter 4 we defined the notion of a "rank function with respect to a strict wellorder relation $\prec$."

EXERCISE 15(PG): (a) Recall Definition 4.34. Assume that W is a wellfounded relation on a set Z, that $\langle X, \prec\rangle$ is a strict w.o., and that $rk : Z \to X$ is a rank function for W with respect to $\prec$. Let $\alpha = ot(\langle X, \prec\rangle)$, and let $F : X \to \alpha$ be the order isomorphism between $\langle X, \prec\rangle$ and $\langle \alpha, \in\rangle$. Show that the composition $F \circ rk : Z \to \alpha$ is a rank function for W with respect to $\in$.

(b) Conclude that for every wellfounded relation there exists a unique rank function with respect to $\in$ whose range is an ordinal.

In the remainder of this text, whenever W is a wellfounded relation on a set Z, the phrase "rank function for W" will refer to the unique rank function with respect to $\in$ which maps Z onto an ordinal.

The next theorem is an important consequence of the fact that **ON** is a proper class.

Theorem 14 (ZF): *For every set x there exists an ordinal $\alpha > 0$ such that x cannot be mapped onto α.*

Proof: The proof rests on the following fact.

Claim 15 (ZF): *Let x be any set. Every function from x onto an ordinal α is the rank function of a strictly wellfounded relation on x.*

Proof: Let x be a set, let $\alpha \in$ **ON**, and let $f : x \to \alpha$ be a function from x onto α. Define a relation R_f on x by: $R_f = \{\langle y, z\rangle : f(y) < f(z)\}$.

We show that R_f is strictly wellfounded: If not, then by Theorem 2.2, there exists a sequence $\langle y_n\rangle_{n\in\omega}$ of elements of x such that $\langle y_{n+1}, y_n\rangle \in R_f$ for all $n \in \omega$. By the definition of R_f, this means that $f(y_{n+1}) \in f(y_n)$ for all $n \in \omega$, which is impossible by the Foundation Axiom.

EXERCISE 16(G): Prove that f is the rank function of R_f.

Exercise 16 concludes the proof of Claim 15.

Now let us return to the proof proper of Theorem 14. Fix any set x. The Power Set Axiom implies that $y = \mathcal{P}(\mathcal{P}(\mathcal{P}(x)))$ is also a set. Recall that every binary relation on x is an element of y.

EXERCISE 17(R): Show that there exists a formula $\psi(r, \alpha)$ of L_S with two free variables such that $\psi(r, \alpha)$ holds if and only if r is a strictly wellfounded relation on x and α is the range of the rank function for r. *Hint:* The formula ψ may contain parameters.

Let ψ be a formula as in Exercise 17, and let $\varphi(r)$ be the formula: $\exists \alpha\, \psi(r, \alpha)$. Then $w = \{r \in y : \varphi(r)\}$ is the set of all strictly wellfounded relations on x. By the instance (A6φ) of the Separation Axiom Schema, w is a set. Moreover, since each strictly wellfounded relation has a unique rank function, $\forall r \in w \exists! \alpha\, \psi(r, \alpha)$. Hence, by the instant (A7ψ) of the Replacement Axiom Schema, the collection $A = \{\alpha : \exists r \in w\, \psi(r, \alpha)\}$ forms a set. By Claim 15, $A = \{\alpha \in \mathbf{ON} : \exists f : x \overset{\text{onto}}{\to} \alpha\}$. Since **ON** is a proper class, there must be an $\alpha \in \mathbf{ON}$ which is not in the set A. This is exactly what Theorem 14 asserts. $\square$

Corollary 16 (ZF): *For every set x there exists an ordinal α such that α cannot be mapped into x by a one-to-one function.*

Proof: If $x = \emptyset$, then no ordinal $\beta > 0$ can be mapped into x. If x is nonempty, let $\alpha > 0$ be an ordinal such that x cannot be mapped onto α. Now Fact 9.17 implies that α cannot be mapped into x by a one-to-one function. $\square$

Now we present two applications of Corollary 16. The following lemma provides the missing direction in the proof of Theorem 9.20.

Lemma 17 (ZF): *If every nonempty set admits a group structure, then every set can be wellordered.*

Proof: Assume that every nonempty set admits a group structure, and let x be any set. We show that x can be wellordered. Let $\alpha > 0$ be an ordinal such that there is no injection from α into x. Consider the set $y = x \cup \alpha$. Since $\alpha > 0$, our assumption implies that y admits a group structure, i.e., there are $e \in y$ and a binary operation $* : y \times y \to y$ such that $\langle y, *, e\rangle$ is a group. Fix such $e, *$.

Claim 18: $\forall z \in x \exists \beta \in \alpha\, (z * \beta \in \alpha)$.

Proof of the claim: Fix $z \in x$, and consider the function $f : \alpha \to y$ given by the formula $f(\beta) = z * \beta$. Since $z^{-1} * f(\beta) = \beta$ for all $\beta \in \alpha$, the function f is one-to-one. By the choice of α, the range of f is not contained in x. But if $f(\beta) \in y \backslash x$, then $f(\beta) \in \alpha$, which proves the claim.

Now let $\leq_a$ be the antilexicographic order on $\alpha \times \alpha$. By Exercise 2.32(b), $\leq_a$ is a wellorder relation. Define a function $g : x \to \alpha \times \alpha$ by letting $g(z)$ be the $\leq_a$-smallest element $\langle \beta, \gamma\rangle$ of $\alpha \times \alpha$ such that $z * \beta = \gamma$. Note that if $g(z) = \langle \beta, \gamma\rangle$, then $z = \gamma * \beta^{-1}$. Hence, g is a one-to-one function. Now define a relation $\preceq$ on x by: $z \preceq z'$ iff $g(z) \leq_a g(z')$. By Exercise 2.41(c), $\preceq$ is a wellorder relation on x. This concludes the proof of Lemma 17. $\square$

Let α be an ordinal. Since $\in$ is a wellfounded relation on α, Theorems 4.21, 4.25, and 4.26 apply. Induction and recursion over $\in$ on ordinals are usually referred to as *transfinite induction* and *transfinite recursion.* If you have mastered the relevant parts of Chapter 4, transfinite induction and transfinite recursion will be nothing new for you. However, you will have to get used to the terminology (some would say "jargon") in which such arguments are usually written.

Suppose you are given an ordinal α, and you want to show that all ordinals $\beta < \alpha$ have a certain property $P(\beta)$. A proof by transfinite induction usually goes like this: "By induction over $\beta < \alpha$. Let $\beta < \alpha$, and suppose that $P(\gamma)$ holds for all $\gamma < \beta$. We show that $P(\beta)$ holds. ..."

Let us recall Theorem 4.21 and see why the above argument really establishes that $P(\beta)$ holds for all $\beta < \alpha$. In the terminology of Theorem 4.21, we want to establish that the set $Y = \{\beta \in \alpha : P(\beta)\}$ is equal to α. For each $\beta \in \alpha$, the initial segment $I_{\in}(\beta)$ is equal to β. Therefore, the last two sentences of the above outline correspond exactly to the implication ($\mathrm{Ind}_{\in}(\beta)$) of Theorem 4.21. Since β was considered to be an arbitrary element of α, by Theorem 4.21, this implication does indeed prove that $Y = \alpha$; i.e., that $P(\beta)$ holds for all $\beta \in \alpha$.

This general pattern of proofs by transfinite induction has several important variations. First of all, since $\in$ wellorders the class **ON**, the class version of Theorem 4.21 applies (compare with Exercise 8.18). Therefore, the same kind of argument can sometimes be used to show that *all* ordinals have a certain property P.

EXERCISE 18(G): Modify the above outline of a proof by transfinite induction so that it would show that $P(\beta)$ holds for all $\beta \in \mathbf{ON}$.

Second, it is often convenient to split the proof of the implication ($\mathrm{Ind}_{\in}(\beta)$) $(\forall \gamma < \beta\, P(\gamma)) \rightarrow P(\beta)$ into three cases: for $\beta = 0$, for β being a successor ordinal, and for β being a limit ordinal > 0.

If $\beta = 0$, then the antecedent of the above implication is vacuously satisfied, and instead of proving an implication, one often has to give a direct proof of $P(0)$.

If $\beta = S(\delta)$ for some δ, then the antecedent of ($\mathrm{Ind}_{\in}(\beta)$) entails in particular that $P(\delta)$ holds, and this is often all one needs for proving $P(\beta)$. Thus, one frequently proves the (*a priori* stronger) implication $P(\delta) \rightarrow P(S(\delta))$ instead of ($\mathrm{Ind}_{\in}(\beta)$).

If β is a limit ordinal > 0, then one has to prove the implication ($\mathrm{Ind}_{\in}(\beta)$). However, in proving this implication, it occasionally helps to use the assumption that β has no largest element.

Transfinite recursion is used to build functions whose domains are ordinals, or functional classes whose domain is the class **ON**. Such functions or functional classes are often called *transfinite sequences* and denoted like this: $\langle x_\beta : \beta < \alpha\rangle$ or $\langle x_\beta : \beta \in \mathbf{ON}\rangle$. Here x_β stands for $F(\beta)$, where F is the function or functional class built by transfinite recursion. In defining a transfinite sequence by transfinite recursion, one usually starts out by specifying x_0 (in the terminology of Theorem 4.25, by defining $G(\emptyset, 0)$), and then one specifies what x_β should be, given the transfinite sequence $\langle x_\gamma : \gamma < \beta\rangle$ (in the terminology of Theorem 4.25, the latter amounts to defining $G(\langle x_\gamma : \gamma < \beta\rangle, \beta)$). Quite often, for successor ordinals $\beta = S(\delta)$, the β-th term x_β of the transfinite sequence depends exclusively on x_δ, whereas for limit ordinals $\beta > 0$, the entire sequence $\langle x_\gamma : \gamma < \beta\rangle$ is used to compute x_β.

EXERCISE 19 (PG): The kind of transfinite recursion we have just described corresponds to Theorem 4.25, the Priciple of Generalized Recursive *Definitions.* What would a transfinite recursion that is based on Theorem 4.26, the Principle of Generalized Recursive *Constructions,* look like?

Now we derive Theorem 9.6 from Corollary 16. Its proof uses both transfinite induction and transfinite recursion. Let us restate the theorem here for your convenience.

Theorem 19 (ZF): *Let X be a set, let $\mathcal{C}$ be a family of subsets of X that is closed under unions of subfamilies that are linearly ordered by inclusion, and let $f : \mathcal{C} \to \mathcal{C}$ be a function with $f(C) \supseteq C$ for all $C \in \mathcal{C}$. Then there exists a $C \in \mathcal{C}$ such that $f(C) = C$.*

Proof: Let $X, \mathcal{C}$, and f be as in the assumption, and let $\alpha > 0$ be an ordinal that cannot be mapped into $\mathcal{C}$ by a one-to-one function. Let us define by transfinite recursion a function $F : \alpha \to \mathcal{P}(X)$ as follows:
Suppose $\beta \in \alpha$ and $F(\gamma)$ has been defined for all $\gamma < \beta$. Define:

$$F(\beta) = \begin{cases} f(\bigcup\{F(\gamma) : \gamma < \beta\}) & \text{if } \bigcup\{F(\gamma) : \gamma < \beta\} \in \mathcal{C}; \\ X & \text{otherwise.} \end{cases}$$

Claim 20: $\forall \gamma < \beta < \alpha\, (F(\gamma) \subseteq F(\beta))$.

Proof: Let $\gamma < \beta < \alpha$. If $F(\beta) = X$, then the inclusion $F(\gamma) \subseteq F(\beta)$ simply follows from the fact that $F(\gamma) \in \mathcal{P}(X)$. If $F(\beta) \neq X$, then $\bigcup\{F(\delta) : \delta < \beta\} \in \mathcal{C}$, and by the assumption on f we have

$$F(\gamma) \subseteq \bigcup\{F(\delta) : \delta < \beta\} \subseteq f(\bigcup\{F(\delta) : \delta < \beta\}) = F(\beta).$$

Claim 21: $F(\beta) \in \mathcal{C}$ *for every* $\beta \in \alpha$.

Proof: By induction over β. Let $\beta < \alpha$, and suppose $F(\gamma) \in \mathcal{C}$ for all $\gamma < \beta$. Consider $\mathcal{F} = \{F(\gamma) : \gamma < \beta\}$. By the inductive assumption, $\mathcal{F} \subseteq \mathcal{C}$. Claim 20 shows that $\mathcal{F}$ is linearly ordered by inclusion. Thus, the assumption on $\mathcal{C}$ implies that $\bigcup \mathcal{F} \in \mathcal{C}$. Since the range of f is contained in $\mathcal{C}$, we conclude that $F(\beta) = f(\bigcup \mathcal{F}) \in \mathcal{C}$.

EXERCISE 20(G): Double-check that the above argument works even if $\beta = 0$, and convince yourself that there was no need to insert a separate clause for $\beta = 0$ in the definition of F.

By Claim 21, F maps α into $\mathcal{C}$. By the choice of α, there are $\gamma < \beta < \alpha$ such that $F(\gamma) = F(\beta)$. Fix such γ, β. If $F(\beta) = X$, then $X \in \mathcal{C}$, and $f(X) = X$ by the assumption on f; so in this case $C = X$ witnesses that the theorem holds. If $F(\beta) \neq X$, then let $C = \bigcup\{F(\delta) : \delta < \beta\}$. The definition of F implies that $C \in \mathcal{C}$. Moreover,

$$F(\beta) = f(\bigcup\{F(\delta) : \delta < \beta\}) \supseteq \bigcup\{F(\delta) : \delta < \beta\} \supseteq F(\gamma) = F(\beta).$$

It follows that $f(C) = C$, and we have proved Theorem 19. □

10.2. Ordinal Arithmetic

Definition 22: Let $\alpha \in \mathbf{ON}$. By recursion over $\beta \in \mathbf{ON}$ we define an ordinal $\alpha + \beta$ as follows:

(i) $\alpha + 0 = \alpha$;
(ii) $\alpha + \beta = S(\alpha + \gamma)$ if $\beta = S(\gamma)$;
(iii) $\alpha + \beta = \bigcup_{\gamma \in \beta}(\alpha + \gamma)$ if β is a limit ordinal > 0.

Lemma 23: *Let $\alpha, \beta \in$ **ON**. Then*

$$\alpha + \beta = \alpha \cup \{\alpha + \gamma : \gamma < \beta\}. \tag{1}$$

Proof: We prove the lemma by induction over β.

If $\beta = 0$, then the right hand side of (1) is $\alpha \cup \emptyset = \alpha$. The left hand side is $\alpha + 0$, which is also equal to α by Definition 22(i).

Assume that $\beta = S(\gamma)$, and that the equation (1) holds for γ. Then

$$\begin{aligned}
&\alpha \cup \{\alpha + \delta : \delta < \beta\} \\
&\quad = \alpha \cup \{\alpha + \delta : \delta < \gamma\} \cup \{\alpha + \gamma\} = \quad \text{(by inductive assumption)} \\
&\quad = \alpha + \gamma \cup \{\alpha + \gamma\} = S(\alpha + \gamma) = \quad \text{(by Definition 22(ii))} \\
&\quad = \alpha + \beta.
\end{aligned}$$

Now assume that β is a limit ordinal and that (1) holds for all $\gamma < \beta$. Then for each $\delta < \beta$ there exists a γ such that $\delta < \gamma < \beta$, and hence:

$$\begin{aligned}
\alpha \cup \{\alpha + \delta : \delta < \beta\} &= \alpha \cup \bigcup_{\gamma < \beta} \{\alpha + \delta : \delta < \gamma\} \\
&= \bigcup_{\gamma < \beta} (\alpha \cup \{\alpha + \delta : \delta < \gamma\}) = \quad \text{(by inductive assumption)} \\
&= \bigcup_{\gamma < \beta} (\alpha + \gamma) = \quad \text{(by Definition 22(iii))} \\
&= \alpha + \beta.
\end{aligned}$$

□

Corollary 24: *Suppose $\alpha, \beta, \gamma, \delta$ are ordinals such that $\gamma < \beta$. Then $\alpha \leq \alpha + \gamma < \alpha + \beta$.*

Proof: This follows immediately from Lemma 23 and Claim 8. □

Corollary 25: *Let $\alpha, \beta \in$ **ON** be such that $\alpha < \beta$. Then there exists exactly one ordinal γ such that $\alpha + \gamma = \beta$.*

Proof: Fix α, β as in the assumptions, and let $F_\alpha(\gamma) = \alpha + \gamma$. It follows from Corollary 24 that F_α is an injective functional class. By Corollary 16, there exists an ordinal δ such that $F_\alpha(\delta) \notin \beta$. Fix such a δ. Since $F_\alpha(\delta) = \alpha + \delta \in$ **ON**, by Lemma 6, we must either have $\alpha + \delta = \beta$, or $\beta \in \alpha + \delta$. In the former case, $\gamma = \delta$ is as required. In the latter case, by Lemma 23, $\beta \in \{\alpha + \gamma : \gamma < \delta\}$, which again yields the desired result. □

Corollary 25 allows us to define a subtraction operation on ordinals: If $\beta \leq \alpha$, then we denote the unique ordinal γ such that $\beta + \gamma = \alpha$ by $\alpha \dot{-} \beta$. If $\alpha < \beta$, then we set $\alpha \dot{-} \beta = 0$.

In Chapter 2 we introduced an operation of addition of arbitrary order types. The next theorem shows that for ordinals, the two notions of addition coincide.

Theorem 26: *Let $\langle X, \preceq_X\rangle$ and $\langle Y, \preceq_Y\rangle$ be w.o.'s, and let $\alpha = ot(\langle X, \preceq_X\rangle)$, $\beta = ot(\langle Y, \preceq_Y\rangle)$. Then $\alpha + \beta = ot(\langle X, \preceq_X\rangle \oplus \langle Y, \preceq_Y\rangle)$.*

Proof: Recall the definition of $\oplus$ from Chapter 2. Let $f : X \to \alpha$ be an order isomorphism between $\langle X, \prec_X\rangle$ and $\langle \alpha, \in\rangle$, and let $g : Y \to \beta$ be an order isomorphism between $\langle Y, \prec_Y\rangle$ and $\langle \beta, \in\rangle$. Define a function $h : X \times \{0\} \cup Y \times \{1\} \to \alpha + \beta$ by:

$$h(x, 0) = f(x); \qquad h(y, 1) = \alpha + g(y).$$

It follows from Lemma 23 that the range of h is $\alpha + \beta$; and Corollary 24 implies that h is order-preserving. By Exercise 2.19, h is an order isomorphism, and we have proved Theorem 26. □

Addition of finite ordinals is the same operation as the usual addition of natural numbers. However, addition of infinite ordinals behaves somewhat differently.

EXERCISE 21(G): Convince yourself that $1 + \omega = \omega$.

Since $\omega + 1 = S(\omega) > \omega$, the result of Exercise 21 implies that addition of ordinals is not commutative. We can also see that the inequality $\alpha < \beta$ does not always imply the inequality $\alpha+\gamma < \beta+\gamma$. However, a weaker form of monotonicity of addition from the left holds.

EXERCISE 22(R): Prove that for all ordinals α, β, γ, if $\alpha \leq \beta$, then $\alpha+\gamma \leq \beta+\gamma$.

Moreover, addition of ordinals is associative.

Claim 27: *If α, β, γ are ordinals, then $(\alpha + \beta) + \gamma = \alpha + (\beta + \gamma)$.*

Proof: This follows immediately from Theorem 26 and Exercise 2.34(a). □

In the remainder of this book, if α is an ordinal, then we shall write $\alpha+1$ instead of $S(\alpha)$, $\alpha+2$ instead of $S(S(\alpha))$, etc. Thus, the next ordinal after ω is $\omega+1$, then come $\omega + 2$, $\omega + 3$, $\omega + 4, \ldots$. By Definition 22(iii), $\bigcup_{n\in\omega}(\omega + n) = \omega + \omega$. Thus, $\omega + \omega$ is the third limit ordinal.[2]

EXERCISE 23(G): What is the fourth limit ordinal? And the 44th?

Well, if you need some practice in writing the Greek letter ω, then the second part of the above exercise is just right for you. But there is a more convenient way of denoting the 44th limit ordinal than putting down 43 ω's set apart by 42 +-signs.

Definition 28: Let $\alpha \in \mathbf{ON}$. By recursion over $\beta \in \mathbf{ON}$ we define an ordinal $\alpha \cdot \beta$ as follows:

(i) $\alpha \cdot 0 = 0$;
(ii) $\alpha \cdot \beta = (\alpha \cdot \gamma) + \alpha$ if $\beta = S(\gamma)$;
(iii) $\alpha \cdot \beta = \bigcup_{\gamma\in\beta}(\alpha \cdot \gamma)$ if β is a limit ordinal > 0.

[2]Recall that the first two limit ordinals are 0 and ω.

In this new notation, $\omega + \omega = \omega \cdot 2$, $\omega + \omega + \omega = \omega \cdot 3$; and the 44th limit ordinal can be written as $\omega \cdot 43$.[3] Multiplication of finite ordinals is the same as the usual multiplication of natural numbers. In the sequel, we shall consider ordinal multiplication as an operation of higher priority than ordinal addition and omit brackets accordingly. In particular, $(\alpha \cdot \beta) + \gamma$ will be written as $\alpha \cdot \beta + \gamma$.

EXERCISE 24(G): (a) What is $2 \cdot \omega$?
(b) Conclude from (a) that ordinal multiplication is not commutative.
(c) Show that there are ordinals α, β, γ such that $(\alpha + \beta) \cdot \gamma \neq \alpha \cdot \gamma + \beta \cdot \gamma$.

The following characterization is the analogue of Lemma 23.

Lemma 29: *Let* $\alpha, \beta \in \mathbf{ON}$. *Then*

(2) $$\alpha \cdot \beta = \{\alpha \cdot \xi + \eta : \xi < \beta \ \wedge \ \eta < \alpha\}.$$

Proof: By induction over β. If $\beta = 0$, then $\alpha \cdot \beta = \alpha \cdot 0 = 0 = \emptyset = \{\alpha \cdot \xi + \eta : \xi < \beta \wedge \eta < \alpha\}$. If (2) holds for γ, and if $\beta = \gamma + 1$, then $\alpha \cdot \beta = \alpha \cdot (\eta + 1) =$ (By Definition 28(ii)) $\alpha \cdot \gamma + \alpha =$ (by Lemma 23) $\alpha \cdot \gamma \cup \{\alpha \cdot \gamma + \eta : \eta < \alpha\} =$ (by inductive assumption) $\{\alpha \cdot \xi + \eta : \xi < \gamma \ \wedge \ \eta < \alpha\} \cup \{\alpha \cdot \gamma + \eta : \eta < \alpha\} = \{\alpha \cdot \xi + \eta : \xi < \beta \ \wedge \ \eta < \alpha\}$.

EXERCISE 25(G): Complete the proof of Lemma 29. □

Corollary 30: *Let* α, β, γ *be ordinals such that* $\alpha > 0$ *and* $\beta < \gamma$. *Then* $\alpha \cdot \beta < \alpha \cdot \gamma$.

EXERCISE 26(G): Prove Corollary 30.

Corollary 31: *Suppose* $\alpha, \beta \in \mathbf{ON}$ *are such that* $1 \leq \alpha < \beta$. *Then there exists a unique pair of ordinals* $\langle \xi, \eta \rangle$ *such that*

(3) $$\eta < \alpha \quad \text{and} \quad \beta = \alpha \cdot \xi + \eta.$$

Proof: Let α, β be as in the assumptions. Existence of ξ and η as in (3) follows immediately from Lemma 29. To show uniqueness, assume that

(4) $$\alpha \cdot \xi + \eta = \alpha \cdot \xi' + \eta' \quad \text{and} \quad \eta, \eta' < \alpha.$$

Case 1: $\xi = \xi'$. Then $\alpha \cdot \xi = \alpha \cdot \xi'$, and it follows from Corollary 24 that $\eta = \eta'$.
Case 2: $\xi \neq \xi'$. Without loss of generality, $\xi < \xi'$. Then $\xi + 1 \leq \xi'$, and $\alpha \cdot \xi + \eta < \alpha \cdot \xi + \alpha = \alpha \cdot (\xi + 1) \leq \alpha \cdot \xi' \leq \alpha \cdot \xi' + \eta'$, which contradicts (4). □

Recall the definition of the antilexicographic order from Chapter 2.

Theorem 32: *Let* $\langle X, \preceq_X \rangle$ *and* $\langle Y, \preceq_Y \rangle$ *be w.o.'s, and let* $\alpha = ot(\langle X, \preceq_X \rangle)$, $\beta = ot(\langle Y, \preceq_Y \rangle)$. *Then* $\alpha \cdot \beta = ot(\langle X, \preceq_X \rangle \otimes^a \langle Y, \preceq_Y \rangle)$.

---

[3] Of course, it would be nicer if $\omega \cdot 44$ were the 44th limit ordinal, but then you would have to say that "0 is the zeroth limit ordinal." Set theorists sometimes talk like this if no outsider is listening.

EXERCISE 27(PG): Prove Theorem 32.

Corollary 33: *Let* $\alpha, \beta, \gamma \in \mathbf{ON}$. *Then*

(a) $(\alpha \cdot \beta) \cdot \gamma = \alpha \cdot (\beta \cdot \gamma)$.
(b) $\alpha \cdot (\beta + \gamma) = \alpha \cdot \beta + \alpha \cdot \gamma$.

Proof: This follows from Theorem 32 and Exercise 2.34. □

The multiplication operation allows us to contemplate more ordinals: The supremum of the set $\{0, \omega \cdot 1, \omega \cdot 2, \omega \cdot 3, \ldots, \omega \cdot n, \ldots\}$ is $\bigcup_{n \in \omega} \omega \cdot n = \omega \cdot \omega$. The next ordinal after that one is $\omega \cdot \omega + 1$, and the next three limit ordinals are $\omega \cdot \omega + \omega$, $\omega \cdot \omega + \omega \cdot 2$, and $\omega \cdot \omega + \omega \cdot 3$. The supremum of the set $\{\omega \cdot \omega + \omega \cdot n : n \in \omega\}$ is $\omega \cdot \omega \cdot 2$, and $\bigcup_{n \in \omega} \omega \cdot \omega \cdot n = \omega \cdot \omega \cdot \omega$. It is time to introduce exponentiation of ordinals.

Definition 34: Let $\alpha \in \mathbf{ON}$. By recursion over β we define an ordinal $\alpha^{\cdot \beta}$ as follows:

(i) $\alpha^{\cdot 0} = 1$;
(ii) $\alpha^{\cdot \beta} = \alpha^{\cdot \gamma} \cdot \alpha$ if $\beta = \gamma + 1$;
(iii) $\alpha^{\cdot \beta} = \bigcup_{\gamma < \beta} \alpha^{\cdot \gamma}$ if β is a limit ordinal > 0.

We write $\alpha^{\cdot \beta}$ in order to distinguish ordinal exponentiation from cardinal exponentiation which will be defined in the next section.

Exponentiation of finite ordinals coincides with the familiar exponentiation of natural numbers (if one is willing to accept the arbitrary convention that $0^0 = 1$). Let us now look at some examples of powers of infinite ordinals. $\omega^{\cdot 1}$ is the same as ω, $\omega^{\cdot 2} = \omega \cdot \omega$, and $\omega^{\cdot 3} = \omega \cdot \omega \cdot \omega$. By Definition 34(iii), $\sup\{\omega^{\cdot n} : n \in \omega\} = \omega^{\cdot \omega}$. For the adventurous among our readers we recommend the following:

EXERCISE 28(X): (a) Visualize the ordinal $\omega^{\cdot (\omega^{\cdot \omega})}$.

(b) Prove that there is a countable ordinal ε such that $\omega^{\cdot \varepsilon} = \varepsilon$. *Hint:* If you have problems showing that the ε you found is countable, you may want to return to this exercise after reading Chapter 11.[4]

If you are not *that* adventurous, then we recommend the following two exercises.

EXERCISE 29(G): (a) Show that $2^{\cdot \omega} = \omega$.

(b) Give an example of ordinals α, β, γ such that $1 < \alpha < \beta$, $1 \leq \gamma$, and $\alpha^{\cdot \gamma} = \beta^{\cdot \gamma}$.

EXERCISE 30(PG): Let $\alpha, \beta, \gamma \in \mathbf{ON}$.
(a) Prove that if $1 < \alpha$ and $\beta < \gamma$, then $\alpha^{\cdot \beta} < \alpha^{\cdot \gamma}$.
(b) Prove that $\alpha^{\cdot (\beta + \gamma)} = \alpha^{\cdot \beta} \cdot \alpha^{\cdot \gamma}$.
(c) Prove that $\alpha^{\cdot (\beta \cdot \gamma)} = (\alpha^{\cdot \beta})^{\cdot \gamma}$.

In the remainder of this section we present two interesting theorems that will not be used later on and can be skipped during the first reading.

[4] The least ordinal ε with $\omega^{\cdot \varepsilon} = \varepsilon$ is denoted by ε_0. It plays an important role in proof theory.

Every natural number has a unique expansion with respect to any base > 1. As the next theorem shows, the same is true for arbitrary ordinals.

Theorem 35 (Cantor's Normal Form Theorem): *Let α, β be ordinals such that $1 < \alpha \leq \beta$. Then there exist a unique k and unique $\gamma_0, \cdots, \gamma_{k-1}$ and $\delta_0, \cdots, \delta_{k-1}$ with $\gamma_0 > \gamma_1 > \cdots > \gamma_{k-1}$ and $0 < \delta_i < \alpha$ for $i < k$ such that*

$$\beta = \alpha^{\cdot \gamma_0} \cdot \delta_0 + \alpha^{\cdot \gamma_1} \cdot \delta_1 + \cdots + \alpha^{\cdot \gamma_{k-1}} \cdot \delta_{k-1}. \tag{5}$$

Proof: Let $\alpha > 1$ be fixed. By induction over $\beta \geq 0$ we show the existence of an expansion as in (5).

If $\beta = 1$, then $\alpha^{\cdot 0} \cdot 1$ is the desired expansion.

Now assume that $\beta > 1$, and that an expansion as in (5) exists for every η such that $1 \leq \eta < \beta$. Let $\Gamma = \{\gamma : \alpha^{\cdot \gamma} \leq \beta\}$.

Claim 36: *The set Γ has a largest element.*

Proof: Suppose not, and let $\gamma_0 = \bigcup \Gamma$. Since the assumptions of Theorem 10(b) are satisfied, γ_0 is a limit ordinal, and $\gamma_0 \notin \Gamma$. On the other hand, by Definition 34(iii),

$$\alpha^{\cdot \gamma_0} = \bigcup \{\alpha^{\cdot \gamma} : \gamma \in \gamma_0\} = \bigcup \{\alpha^{\cdot \gamma} : \gamma \in \Gamma\} \subseteq \beta.$$

By Claim 8, $\alpha^{\cdot \gamma_0} \leq \beta$, and thus $\gamma_0 \in \Gamma$. We have reached a contradiction.

Now let γ_0 be the largest ordinal in Γ. By Corollary 31, there exist ordinals $\delta_0 > 0$ and $\beta_1 < \beta$ such that $\beta = \alpha^{\cdot \gamma_0} \cdot \delta_0 + \beta_1$. Note that $\delta_0 < \alpha$: If not, then $\beta = \alpha^{\cdot \gamma_0} \cdot \delta_0 + \beta_1 \geq \alpha^{\cdot \gamma_0} \cdot \delta_0 \geq \alpha^{\cdot \gamma_0} \cdot \alpha = \alpha^{\cdot (\gamma_0 + 1)}$, which contradicts the choice of γ_0.

Now, if $\beta_1 = 0$, then $\alpha^{\cdot \gamma_0} \cdot \delta_0$ is the desired expansion. If $\beta_1 > 0$, then, by the inductive assumption, β_1 has an expansion as in (5); say, $\beta_1 = \alpha^{\cdot \gamma_1} \cdot \delta_1 + \cdots + \alpha^{\cdot \gamma_{k-1}} \cdot \delta_{k-1}$ with $\gamma_1 > \gamma_2 > \cdots > \gamma_{k-1}$ and $0 < \delta_i < \alpha$ for $0 < i < k$.

EXERCISE 31(PG): Show that, in the above notation, $\gamma_0 > \gamma_1$.

Thus, adding the term $\alpha^{\cdot \gamma_0} \cdot \delta_0$ from the left to the expansion of β_1 yields the desired expansion for β.

Now we show uniqueness. Suppose towards a contradiction that β is a counterexample, and let:

$$\beta = \alpha^{\cdot \gamma_0} \cdot \delta_0 + \alpha^{\cdot \gamma_1} \cdot \delta_1 + \cdots + \alpha^{\cdot \gamma_{k-1}} \cdot \delta_{k-1},$$
$$\beta = \alpha^{\cdot \gamma_0} \cdot \delta_0' + \alpha^{\cdot \gamma_1} \cdot \delta_1' + \cdots + \alpha^{\cdot \gamma_{k-1}} \cdot \delta_{k-1}'$$

be two different expansions of β such that $\gamma_0 > \gamma_1 > \cdots > \gamma_{k-1}$, $\delta_i, \delta_i' < \alpha$ for $i < k$, and $\max\{\delta_0, \delta_0'\} > 0$. (By allowing $\delta_i, \delta_i' = 0$, we may conveniently assume that the sequence of exponents is the same for both expansions of β.) Moreover, if $\delta_0 = \delta_0'$, then $\beta_1 = \alpha^{\cdot \gamma_1} \cdot \delta_1' + \cdots + \alpha^{\cdot \gamma_{k-1}} \cdot \delta_{k-1}'$ is a smaller ordinal with two different expansions. But the latter contradicts the choice of β, so we must have $\delta_0 \neq \delta_0'$. Without loss of generality, assume $\delta_0 < \delta_0'$. But then:

$$\begin{aligned}
\beta &= \alpha^{\cdot\gamma_0} \cdot \delta_0 + \alpha^{\cdot\gamma_1} \cdot \delta_1 + \cdots + \alpha^{\cdot\gamma_{k-1}} \cdot \delta_{k-1} \\
&< \alpha^{\cdot\gamma_0} \cdot \delta_0 + \alpha^{\cdot\gamma_1} \cdot \delta_1 + \cdots + \alpha^{\cdot\gamma_{k-1}} \cdot \alpha \\
&= \alpha^{\cdot\gamma_0} \cdot \delta_0 + \alpha^{\cdot\gamma_1} \cdot \delta_1 + \cdots + \alpha^{\cdot(\gamma_{k-1}+1)} \\
&\leq \alpha^{\cdot\gamma_0} \cdot \delta_0 + \alpha^{\cdot\gamma_1} \cdot \delta_1 + \cdots + \alpha^{\cdot\gamma_{k-2}} \cdot \delta_{k-2} + \alpha^{\cdot\gamma_{k-2}} \\
&= \alpha^{\cdot\gamma_0} \cdot \delta_0 + \alpha^{\cdot\gamma_1} \cdot \delta_1 + \cdots + \alpha^{\cdot\gamma_{k-2}} \cdot (\delta_{k-2} + 1) \\
&\leq \alpha^{\cdot\gamma_0} \cdot \delta_0 + \alpha^{\cdot\gamma_1} \cdot \delta_1 + \cdots + \alpha^{\cdot\gamma_{k-2}} \cdot \alpha \\
&\leq \cdots \leq \alpha^{\cdot\gamma_0} \cdot (\delta_0 + 1) \leq \alpha^{\cdot\gamma_0} \cdot \delta_0' \\
&\leq \alpha^{\cdot\gamma_0} \cdot \delta_0' + \alpha^{\cdot\gamma_1} \cdot \delta_1' + \cdots + \alpha^{\cdot\gamma_{k-1}} \cdot \delta_{k-1}' = \beta,
\end{aligned}$$

a contradiction. □

We conclude this section with the proof of Goodstein's Theorem, which is an application of infinite ordinal arithmetic to number theory. Before we can formulate the theorem, we need to describe the expansion of a natural number in superbase n. Let us begin with an example. If we expand the number 27 in base 2 we get:

$$27 = 2^4 + 2^3 + 2^1 + 2^0.$$

This expansion still contains a four and a three which we would like to eliminate. We can do this by expanding these two numbers in base 2 as well:

$$27 = 2^{(2^2)} + 2^{(2^1+1)} + 2^1 + 2^0.$$

For larger numbers we may have to repeat this procedure more often. For example, the process of expanding the number 521 in base 2 and eliminating all exponents larger than 2 yields:

$$521 = 2^9 + 2^3 + 1 = 2^{(2^3+1)} + 2^{(2+1)} + 1 = 2^{(2^{(2+1)}+1)} + 2^{(2+1)} + 1.$$

This representation is called *expansion in superbase* 2. Similar expansions in superbase n exist for every natural number $n > 1$. For example, expanding 521 in superbase 3 we get:

$$521 = 2 \cdot 3^5 + 3^3 + 2 \cdot 3 + 2 = 2 \cdot 3^{(3+2)} + 3^3 + 2 \cdot 3 + 2.$$

Now let n be an arbitrary natural number ≥ 2. We want to describe a function S_n that assigns to a natural number k the number one obtains by expanding k in superbase n and replacing all occurrences of n in this expansion by $n+1$. Here is a formal definition of S_n:

Definition 37: For each $n < \omega$ with $n \geq 2$, let $S_n : \omega \to \omega$ be the function defined recursively by:

$$S_n(k) = k \quad \text{for } k < n;$$
$$S_n(k \cdot n^t + b) = k(n+1)^{S_n(t)} + S_n(b) \quad \text{for } k < n,\ b < n^t,\ t \geq 1.$$

Let us consider an example:

$$\begin{aligned}
S_3(125) &= S_3(3^4 + 44) = 4^{S_3(4)} + S_3(44) = 4^{S_3(3+1)} + S_3(44) \\
&= 4^{4+1} + S_3(44) = 4^5 + S_3(3^3 + 17) = 4^5 + 4^{S_3(3)} + S_3(9+8) \\
&= 4^5 + 4^4 + 4^2 + S_3(2 \cdot 3 + 2) = 4^5 + 4^4 + 4^2 + 2 \cdot 4 + 2 = 1306.
\end{aligned}$$

Now we define functions f_n that assign to each natural number a countable ordinal.

Definition 38: For each $n < \omega$ with $n \geq 2$, let $f_n : \omega \to \omega_1$ be the function defined recursively by:

$$f_n(k) = k \quad \text{for } k < n;$$
$$f_n(k \cdot n^t + b) = \omega^{f_n(t)} \cdot k + f_n(b) \quad \text{for } k < n,\ b < n^t,\ t \geq 1.$$

Roughly speaking, $f_n(k)$ is the ordinal obtained by replacing in the expansion of k in superbase n all occurrences of n by ω (and reversing the order of the factors). Thus, for example, $f_3(521) = \omega^{(\omega+2)} \cdot 2 + \omega^{\omega} + \omega \cdot 2 + 2$.

EXERCISE 32(R): (a) Show that if $k < m$ then $f_n(k) < f_n(m)$.
(b) Show that for all $n, k < \omega$, $n \geq 2$ the equality $f_{n+1}(S_n(k)) = f_n(k)$ holds.

Definition 39: For each $n < \omega$ with $n \geq 1$ let $g_n : \omega \to \omega$ be given by

$$g_1(m) = m;$$

$$g_{n+1}(m) = \begin{cases} S_{n+1}(g_n(m)) - 1, & \text{if } g_n(m) > 0; \\ 0, & \text{otherwise.} \end{cases}$$

Theorem 40(Goodstein): *For each m there exists an $n \geq 1$ such that $g_n(m) = 0$.*

Before proving Theorem 40, we illustrate by means of an example how the sequences $(g_n(m))_{1 \leq n < \omega}$ behave. Let $m = 13$. Then we have

$g_1(13) = 13;$	$13 = 2^3 + 2^2 + 1 = 2^{2+1} + 2^2 + 1$
	$S_2(13) = 3^4 + 3^3 + 1 = 109$
$g_2(13) = 108;$	$108 = 3^4 + 3^3$
	$S_3(108) = 4^5 + 4^4 = 1280$
$g_3(13) = 1279;$	$1279 = 4^{4+1} + 3 \cdot 4^3 + 3 \cdot 4^2 + 3 \cdot 4 + 3$
	$S_4(1279) = 5^6 + 3 \cdot 5^3 + 3 \cdot 5^2 + 3 \cdot 5 + 3 = 16093$
$g_4(13) = 16092;$	$16092 = 5^{5+1} + 3 \cdot 5^3 + 3 \cdot 5^2 + 3 \cdot 5 + 2$
	$S_5(16092) = 6^7 + 3 \cdot 6^3 + 3 \cdot 6^2 + 3 \cdot 6 + 2 = 280712$

The next few terms of the sequence are 280711, 5765998, 134219591, 3486786855, 100000003325.
Considering how rapidly the sequence increases initially, it is rather surprising that it eventually attains the value 0.

EXERCISE 33 (G): Calculate $g_i(4)$ for $i = 1, \ldots, 12$.

Proof of Theorem 40: Let $m < \omega$ be given. Consider the sequence

$$(f_{n+1}(g_n(m)))_{1 \leq n < \omega}.$$

As long as $g_{n+1}(m) > 0$, according to the definition of $g_{n+1}(m)$, we have

$$f_{n+2}(g_{n+1}(m)) = f_{n+2}(S_{n+1}(g_n(m)) - 1).$$

Exercise 32(a) implies that

$$f_{n+2}(S_{n+1}(g_n(m) - 1)) < f_{n+2}(S_{n+1}(g_n(m))),$$

and Exercise 32(b) yields

$$f_{n+2}(S_{n+1}(g_n(m))) = f_{n+1}(g_n(m)).$$

Thus

$$f_{n+2}(g_{n+1}(m)) < f_{n+1}(g_n(m)).$$

Since there is no infinite decreasing sequence of ordinals, for sufficiently large n we must have $g_n(m) = 0$. □

Mathographical Remark

Theorem 40 was proved in 1944 by R. L. Goodstein. Since the theorem itself concerns only finite numbers, mathematicians were trying for a long time to find a proof that would not make reference to infinite sets. Finally, L. Kirby und J. Paris showed in their paper *Accessible independence results for Peano arithmetic,* Bull. London Math. Soc. 14 (1982), 285-293, that Goodstein's Theorem is not provable in Peano Arithmetic.[5] This result implies that *any* proof of Goodstein's Theorem must use the notion of an infinite set.

[5] Assuming, of course, that Peano Arithmetic is a consistent theory.

CHAPTER 11

The Cardinals

11.1. Initial Ordinals

In Chapter 3 we announced that we shall eventually pick one set out of every equivalence class of the equivalence $\approx$ and call it a cardinal number.

Definition 1: Let X be a set. We define the *cardinal number of* X to be the smallest ordinal κ such that $X \approx \kappa$. This ordinal will be denoted by $|X|$.

The above definition suggests that some, but not all ordinals are cardinals. If α is an ordinal such that $\alpha \approx \beta$ for some $\beta < \alpha$, then α is not a cardinal. For example, none of the ordinals $\omega + 4$, $\omega \cdot 4$, $\omega^{\cdot 4}$ is a cardinal.

EXERCISE 1(G): Convince yourself that each of the ordinals mentioned above is equipotent with the smaller ordinal ω.

An ordinal α that is not equipotent with any smaller ordinal will also be called an *initial ordinal.* Thus cardinals are initial ordinals. While ordinals in general will be denoted by lower case Greek letters, we shall reserve the letters $\kappa, \lambda, \mu, \nu$ for initial ordinals.

EXERCISE 2(PG): Prove that for every set X there exists exactly one initial ordinal κ such that $X \approx \kappa$.

It is not possible to prove in ZF that every set is equipotent with an initial ordinal.

Theorem 2 (ZF): *Suppose every set is equipotent with some initial ordinal. Then the Axiom of Choice holds.*

EXERCISE 3(PG): (a) Prove Theorem 2.
(b) Prove that condition (SC) of Theorem 9.19 implies the Wellorder Principle (WO). *Hint:* Use Theorem 10.14.

Let us contemplate the first few initial ordinals. By default, 0 is an initial ordinal. More generally, if n is any natural number, then by Theorem 4.8, $n \not\approx m$ for any $m < n$.

EXERCISE 4(G): Look up 4.8 and convince yourself that what we just wrote is true.

The first infinite ordinal ω is not equipotent with any smaller ordinal, since all smaller ordinals are finite. Thus, ω is the first infinite cardinal. As Exercise 1 shows, ω is followed by a large number of ordinals which are *not* cardinals. Nevertheless, it is not true that all ordinals are countable: By Theorem 3.5, $|\mathcal{P}(\omega)| > \omega$. Hence, there must be a $\kappa > \omega$ such that $|\mathcal{P}(\omega)| = \kappa$. We do not know whether this κ is the next cardinal after ω; but by wellfoundedness of **ON**, there must be a smallest uncountable ordinal. This first uncountable ordinal is denoted by ω_1.

EXERCISE 5(G): (a) Show that if α is any ordinal, then there exists an initial ordinal $\lambda > \alpha$.

(b) Infer that the class **Card** $= \{\lambda \in \mathbf{ON} : \lambda \text{ is a cardinal}\}$ is proper. *Hint:* Use Theorem 10.13.

For every cardinal κ, the smallest cardinal that exceeds κ will be called the *cardinal successor of* κ and denoted by κ^+. Thus, $6^+ = 7$, and $\omega^+ = \omega_1$. The cardinal successor of ω_1 is denoted by ω_2; ω_2^+ is denoted by ω_3, and so on. Instead of $(\kappa^+)^+$ we write κ^{++}. If $\lambda = \kappa^+$ for some κ, then λ is called a *successor cardinal.* If a cardinal λ is not of the form κ^+ for any κ, then λ is called a *limit cardinal.* 0 and ω are limit cardinals, whereas all positive finite cardinals and the cardinals $\omega_1, \omega_2, \omega_3, \ldots$ are successor cardinals. We shall prove later in this chapter that all infinite cardinals are limit ordinals. Thus, e.g. ω_1 is simultaneously a limit ordinal and a successor cardinal.

The next theorem provides a method for constructing more limit cardinals.

Theorem 3: *Let A be a set of cardinals. Then $\bigcup A$ is also a cardinal. Moreover, if A does not contain a largest element, then $\bigcup A$ is a limit cardinal.*

Proof: Let A be as in the assumptions, and denote $\bigcup A$ by α. By Theorem 10.10, α is an ordinal.

Let us assume that α is not a cardinal, and let $\kappa = |\alpha|$. Then $\kappa < \alpha$. Since $\bigcup A$ is not contained in κ, there exists a cardinal $\lambda \in A$ with $\kappa < \lambda$. But then we have $|\alpha| = \kappa < \lambda \leq \alpha$, which is impossible by the following observation.

EXERCISE 6(G): Show that if α is an ordinal, and $\kappa = |\alpha| < \alpha$, then no ordinal β such that $\kappa < \beta \leq \alpha$ is a cardinal.

Now we show that if A has no largest element, then $\alpha = \bigcup A$ is a limit cardinal. Assume towards a contradiction that $\alpha = \kappa^+$ for some κ. Then A must contain an element λ such that $\kappa < \lambda$. If A has no largest element, there is also a cardinal $\mu \in A$ such that $\lambda < \mu$. For these choices of λ and μ we have $\lambda \geq \kappa^+$ and $\mu > \kappa^+ = \alpha$, which is impossible, since by Theorem 10.10(b) α exceeds every element of A. □

EXERCISE 7(PG): Show that the collection of all limit cardinals is a proper class.

In many applications of cardinals, it is convenient to have an enumeration of all infinite cardinals. This enumeration can be defined by transfinite recursion as follows:

(i) $F(0) = \omega$;
(ii) $F(\alpha + 1) = F(\alpha)^+$;
(iii) $F(\alpha) = \bigcup_{\beta<\alpha} F(\beta)$ if α is a limit ordinal > 0.

If $\alpha > 0$, it is customary to denote $F(\alpha)$ by ω_α.

EXERCISE 8(PG): (a) Convince yourself that clauses (i)–(iii) uniquely determine a functional class F with domain **ON** whose values are infinite cardinals.

(b) Convince yourself that if $\alpha < \beta$, then $\omega_\alpha < \omega_\beta$. *Hint:* Fix α and use transfinite induction over $\beta > \alpha$.

(c) Show that $\omega_\alpha \geq \alpha$ for every $\alpha \in \mathbf{ON}$.

(d) Show that the range of F is the class of all infinite cardinals.

(e) Show that ω_α is a limit cardinal iff α is a limit ordinal.

We conclude this section with a typical application of initial ordinals. Let X be a topological space. Recall from Section 9.2 that a set $B \subseteq X$ is a *Bernstein set* if $K \cap B \neq \emptyset$ and $K \backslash B \neq \emptyset$ for every uncountable closed subset K of X. We say that X is a *Polish space* if X is separable and completely metrizable.

Theorem 4: *Let X be a Polish space. Then there exists a Bernstein set in X.*

Proof: We need two facts about Polish spaces.

Claim 5: *Every closed subset of a Polish space is either countable or of cardinality $2^{\aleph_0}$.*

The proof of Claim 5 will be given in Chapter 14 of Volume II.

Claim 6: *Let X be a Polish space, and let $\mathbf{K}(X)$ be the family of all uncountable closed subsets of X. Then $|\mathbf{K}(X)| \leq 2^{\aleph_0}$.*

Proof: Since every Polish space has countable weight, Claim 6 is an immediate consequence of the following exercise.

EXERCISE 9(G): Show that if Y is a topological space of weight λ, then there are at most 2^λ closed sets in X.

Now let X be a Polish space. If X is countable, then every subset of X is a Bernstein set; so assume X is uncountable. Since X is a closed subset of itself, by Claim 5, $|X| = 2^{\aleph_0}$. Let $\kappa = 2^{\aleph_0}$. By Claim 6, there exists a surjection $f : \kappa \to \mathbf{K}(X)$. In other words, there exists a transfinite sequence $\overline{K} = \langle K_\xi : \xi < \kappa \rangle$ such that every uncountable closed subset K of X appears at least once in this sequence. We fix such a transfinite sequence $\overline{K}$, and construct by recursion over $\xi < \kappa$ transfinite sequences $\bar{b} = \langle b_\xi : \xi < \kappa \rangle$ and $\bar{c} = \langle c_\xi : \xi < \kappa \rangle$ such that for all $\eta \leq \xi < \kappa$:
(i) $\{b_\xi, c_\xi\} \subset K_\xi$;
(ii) $c_\eta \neq b_\xi$;
(iii) $b_\eta \neq c_\xi$.

Since we are doing a recursive *construction*, it suffices to show that at each stage of the construction we can find b_ξ, c_ξ satisfying (i)–(iii).

So assume we are at stage ξ of the construction. Since the set $\{c_\eta : \eta < \xi\}$ is of cardinality $\leq |\xi| < \kappa$, and since $|K_\xi| = \kappa$, we can find $b_\xi \in K_\xi$ such that $b_\xi \neq c_\eta$ for every $\eta < \xi$. Having chosen b_ξ, let us consider the set $\{b_\eta : \eta \leq \xi\}$. This set has cardinality at most $|\xi + 1|$.

EXERCISE 10(G): Show that $|\xi + 1|$ is either finite, or equal to $|\xi|$.

It follows from Exercise 10 that $|\{b_\eta : \eta \leq \xi\}| < \kappa$. Hence, there exists $c_\xi \in K_\xi \backslash \{b_\eta : \eta \leq \xi\}$. We pick such c_ξ and proceed to the next stage of the recursive construction.

EXERCISE 11(PG): Find a function G^* which shows that the above construction is an application of Theorem 4.26.

EXERCISE 12(G): Convince yourself that the sequences $\langle b_\xi : \xi < \kappa \rangle$ and $\langle c_\xi : \xi < \kappa \rangle$ constructed above satisfy conditions (i)–(iii).

Now let $B = \{b_\xi : \xi < \kappa\}$ and $C = \{c_\xi : \xi < \kappa\}$. By (ii) and (iii), $B \cap C = \emptyset$. Moreover, since every uncountable closed subset K of X appears as some K_ξ in the sequence $\overline{K}$, condition (i) implies that $K \cap B \neq \emptyset$ and $K \cap C \neq \emptyset$ for every uncountable closed subset of X. Since $B \cap C = \emptyset$, this implies that $K \backslash B \neq \emptyset$. Therefore, B is a Bernstein set (and so is C, by the way). □

11.2. Cardinal Arithmetic

In Chapter 3, we already defined addition, multiplication, and exponentiation of cardinals. Let us reformulate these definitions in the terminology of cardinal numbers.

Definition 7: Let κ, λ be cardinals. We define:
$\kappa + \lambda = |\kappa \times \{0\} \cup \lambda \times \{1\}|$;
$\kappa \cdot \lambda = |\kappa \times \lambda|$;
$\kappa^\lambda = |{}^\lambda \kappa|$.

Unfortunately, in the set-theoretic literature, the same symbols + and are used for both cardinal and ordinal addition and multiplication. Since this may at times be confusing, we shall often explicitly state whether a particular addition/multiplication operation is supposed to be cardinal or ordinal addition/multiplication. Another device that helps eliminate ambiguity is to denote the α-th infinite cardinal by $\aleph_\alpha$ instead of ω_α. Thus, when we write $\aleph_\alpha \cdot \aleph_\beta$, we refer to cardinal multiplication, whereas $\omega_\alpha \cdot \omega_\beta$ refers to ordinal multiplication. While we shall stick to this convention throughout this text, it is not universally accepted in the literature. Many authors write $\omega_\alpha \cdot \omega_\beta$ when they want to refer to multiplication of cardinals.

Our first goal in this section is to prove Theorem 3.19. We split this into two parts.

Theorem 8: *Let κ, λ be cardinals such that $\kappa \leq \lambda$ and λ is infinite. Then $\kappa + \lambda = \lambda$.*

Proof: Earlier in this section, we already proved Theorem 8 for a special case: Exercise 10 shows that $1 + \lambda = \lambda$ for every infinite cardinal λ. This observation has an important corollary.

Lemma 9: *Every infinite cardinal is a limit ordinal.*

EXERCISE 13(G): Derive Lemma 9 from the result of Exercise 10.

We need one more basic property of ordinal addition.

Lemma 10: *If α is any ordinal, then there exist a unique limit ordinal β and a unique $n \in \omega$ such that $\alpha = \beta + n$.*

Proof: Let $\alpha \in \mathbf{ON}$. Recursively define a sequence $\langle \alpha_k : k \in \omega \rangle$ as follows: First, set $\alpha_0 = \alpha$. Then suppose α_n has already been defined. If α_n is a successor ordinal, then let α_{n+1} be the unique ordinal β such that $\alpha_n = \beta + 1$. If α_n is a limit ordinal, then let $\alpha_{n+1} = \alpha_n$.

By Theorem 2.2 and wellfoundedness of $\in$, it is not possible that $\alpha_{n+1} \in \alpha_n$ for all $n \in \omega$. Therefore, the equality $\alpha_n = \alpha_{n+1}$ must hold for some $n \in \omega$. Let n_0 be the smallest such n.

EXERCISE 14(G): Convince yourself that α_{n_0} is a limit ordinal, and that $\alpha = \alpha_{n_0} + n_0$. *Hint:* For a formal proof, use induction over n_0.

EXERCISE 15(G): Show that if $\beta, \beta' \in \mathbf{Lim}$ and $n, n' \in \omega$ are such that $\alpha = \beta + n = \beta' + n'$, then $\beta = \beta'$ and $n = n'$.

Now let us return to the proof of Theorem 8. Let κ, λ be as in the assumptions. By Theorem 3.13(f1), $\lambda = 0 + \lambda \leq \kappa + \lambda \leq \lambda + \lambda$. Therefore, by the Cantor-Schröder-Bernstein Theorem, it suffices to prove that $\lambda + \lambda \leq \lambda$.

By Lemma 9, λ is a limit ordinal. In particular, if $\alpha \in \lambda$ and $m \in \omega$, then $\alpha + m \in \lambda$. We define a function $f : \lambda \times \{0\} \cup \lambda \times \{1\} \to \lambda$ as follows: Suppose $\alpha \in \lambda$, $\beta \in \mathbf{Lim}$, and $n \in \omega$ are such that $\alpha = \beta + n$. Then we let $f(\alpha, i) = \beta + 2n + i$, where $i \in \{0, 1\}$. The following exercise concludes the proof of Theorem 8.

EXERCISE 16(G): Convince yourself that f is a well defined, one-to-one function from $\lambda \times \{0\} \cup \lambda \times \{1\}$ into λ. □

Theorem 11 (Hessenberg): *Let κ, λ be cardinals such that $1 \leq \kappa \leq \lambda$ and λ is infinite. Then $\kappa \cdot \lambda = \lambda$.*

Proof: Again, it suffices to prove that $\lambda \cdot \lambda \leq \lambda$ for every infinite cardinal λ.

EXERCISE 17(G): Why?

We prove this by induction over infinite cardinals λ.

We have already shown that $\aleph_0 \cdot \aleph_0 = \aleph_0$, but the present proof subsumes this special case. All we need is the inductive assumption that λ is an infinite cardinal, and the fact that $\kappa \cdot \kappa < \lambda$ for every cardinal $\kappa < \lambda$. The latter is of course true if $\lambda = \aleph_0$. We define a strict wellorder relation $<_w$ on $\lambda \times \lambda$ as follows:

$$\langle \alpha, \beta \rangle <_w \langle \gamma, \delta \rangle \quad \text{iff} \quad \alpha + \beta < \gamma + \delta \text{ or}$$
$$\alpha + \beta = \gamma + \delta \text{ and } \alpha < \gamma$$

EXERCISE 18(G): Convince yourself that $\langle \lambda \times \lambda, <_w \rangle$ is a strict l.o.

We show that $<_w$ is a wellfounded relation. Let A be a nonempty subset of $\lambda \times \lambda$, and let $\gamma^* = min\{\alpha + \beta : \langle \alpha, \beta \rangle \in A\}$. Consider $B = \{\langle \alpha, \beta \rangle \in A : \alpha + \beta = \gamma^*\}$, and let $\alpha^* = min\{\alpha : \exists \beta\, (\langle \alpha, \beta \rangle \in B)\}$. By Corollary 10.25, there exists a unique β^* such that $\alpha^* + \beta^* = \gamma^*$. The pair $\langle \alpha^*, \beta^* \rangle$ is the smallest element of A.

Now let $\delta = ot(\langle \lambda \times \lambda, <_w \rangle)$. Since $|\delta| = |\lambda \times \lambda|$, it suffices to prove that $\delta \leq \lambda$.

EXERCISE 19(G): Show that λ is the largest ordinal γ with the property that every proper initial segment of γ has cardinality less than λ.

By the above exercise, it suffices to prove the following.

Claim 12: *Every proper initial segment of the strict w.o. $\langle \lambda \times \lambda, <_w \rangle$ has cardinality less than λ.*

Proof: Let $\beta, \gamma \in \lambda$, and let $I(\langle \beta, \gamma \rangle) = \{\langle \beta', \gamma' \rangle : \langle \beta', \gamma' \rangle <_w \langle \beta, \gamma \rangle\}$ be a proper initial segment of $\langle \lambda \times \lambda, <_w \rangle$. Let α be an ordinal such that $\beta + \gamma < \alpha < \lambda$.

EXERCISE 20(G): Show that such an α exists.

By the choice of α, the inclusion $I(\langle \beta, \gamma \rangle) \subseteq \{\langle \beta', \gamma' \rangle : \beta', \gamma' < \alpha\}$ holds. But the right hand side of the above inclusion is equal to $\alpha \times \alpha$, and by the inductive assumption, $|\alpha \times \alpha| < \lambda$. This concludes the proof of Claim 12, and simultaneously, of Theorem 11. □

Cardinal addition and multiplication can also be defined for infinite sets of cardinals. If I is a set, and if $\{\kappa_i : i \in I\}$ is a set of cardinals, then we define

$$\sum_{i \in I} \kappa_i = \left| \bigcup_{i \in I} \kappa_i \times \{i\} \right|.$$

In this terminology, Theorem 3.17(b) can be written as follows:

Fact 13: *If $|I| = \aleph_0$ and if $\kappa_i \leq \aleph_0$ for every $i \in I$, then*

$$\sum_{i \in I} \kappa_i \leq \aleph_0.$$

This fact generalizes as follows:

Theorem 14: *Let κ be an infinite cardinal, let $|I| \leq \kappa$; and for each $i \in I$, let κ_i be a cardinal such that $\kappa_i \leq \kappa$. Then $\sum_{i\in I} \kappa_i \leq \kappa$.*

Proof: For each $i \in I$, let $f_i : \kappa_i \to \kappa$ be a one-to-one function, and let $f : I \to \kappa$ also be one-to-one. Define a function $F : \bigcup_{i\in I} \kappa_i \times \{i\} \to \kappa \times \kappa$ by letting $F(\xi, i) = \langle f_i(\xi), f(i)\rangle$.

EXERCISE 21(G): Convince yourself that F is a one-to-one function. □

Since $|\kappa \times \kappa| = \kappa$, Theorem 14 follows.

Corollary 15: *If λ is an infinite successor cardinal, then λ is not a union of fewer than λ sets, each of which has cardinality less than λ.*

If $\langle \mu_i : i \in I\rangle$ is an enumeration of all cardinals that are smaller than a fixed cardinal λ, and if κ is fixed, then we write $\kappa^{<\lambda}$ instead of $\sum_{i\in I} \kappa^{\mu_i}$.

EXERCISE 22(G): Show that condition (FS) of Theorem 9.19 is a theorem of ZFC, i.e., prove that $\kappa^{<\aleph_0} = \kappa$ for every infinite cardinal κ.

The property expressed in Corollary 15 distinguishes infinite successor cardinals from such cardinals as $\aleph_\omega$. (Note that $\aleph_\omega = \bigcup_{i\in\omega} \aleph_i$.) It deserves a name.

Definition 16: (a) Let α be an ordinal, and let $G \subseteq \alpha$. We say that G is *cofinal in* α if $\forall\beta \in \alpha\exists\gamma \in G\ \gamma \geq \beta$. The *cofinality of* α, denoted by $cf(\alpha)$, is the smallest ordinal δ such that there exists a function $f : \delta \to \alpha$ whose range is cofinal in α.

(b) Let κ be a cardinal. We say that κ is *regular* if $cf(\kappa) = \kappa$; otherwise we say that κ is *singular.*

EXERCISE 23(PG): (a) Show that α is a successor ordinal iff $cf(\alpha) = 1$.

(b) Show that for every ordinal α,

$$cf(\alpha) = min\{|G| : G \text{ is a cofinal subset of } \alpha\}.$$

(c) Show that for every ordinal α, the cofinality of α is an initial ordinal.

(d) Show that $cf(cf(\alpha)) = cf(\alpha)$ for every ordinal α.

(e) Conclude that for every $\alpha \in \mathbf{ON}$, $cf(\alpha)$ is a regular cardinal.

(f) Show that if $cf(\alpha) = \delta$, then there exists a *strictly increasing* function $f : \delta \to \alpha$ whose range is cofinal in α.

Lemma 17: *Let κ be an infinite cardinal. Then κ is singular iff there exist an $\alpha < \kappa$ and a set of cardinals $\{\kappa_\xi : \xi < \alpha\}$ such that $\kappa_\xi < \kappa$ for every $\xi < \alpha$, and $\kappa = \sum_{\xi<\alpha} \kappa_\xi$.*

Proof: Suppose $cf(\kappa) = \alpha < \kappa$, and let $f : \alpha \to \kappa$ be a strictly increasing function whose range is cofinal in κ. For $\xi < \alpha$, let $\kappa_\xi = |f(\xi)\backslash \bigcup_{\eta<\xi} f(\eta)|$.

EXERCISE 24(G): Show that $\sum_{\xi<\alpha} \kappa_\xi = \kappa$.

Now assume that $\sum_{\xi<\alpha}\kappa_\xi = \kappa$ for some $\alpha < \kappa$ and $\kappa_\xi < \kappa$ for all $\xi < \alpha$. Without loss of generality, we may assume that for all $\beta < \alpha$ the inequality $\sum_{\xi<\beta}\kappa_\xi < \kappa$ holds.

EXERCISE 25(G): Why?

Fix a bijection $F : \bigcup_{\xi<\alpha}\kappa_\xi \times \{\xi\} \to \kappa$, and consider two cases:

Case 1: There exists a $\beta < \alpha$ such that $F[\bigcup_{\xi<\beta}\kappa_\xi \times \{\xi\}]$ is cofinal in κ. Then $cf(\kappa) \leq |\bigcup_{\xi<\beta}\kappa_\xi \times \{\xi\}| = \sum_{\xi<\beta}\kappa_\xi < \kappa$.

Case 2: The assumption of Case 1 does not hold. Define a function $f : \alpha \to \kappa$ by the formula $f(\beta) = \sup\{F[\bigcup_{\xi<\beta}\kappa_\xi \times \{\xi\}]\}$. The assumption of Case 2 implies that f maps α into κ. Since α is a limit ordinal,

EXERCISE 26(G): Why?

the range of f must be cofinal in κ: For each $\gamma \in \kappa$, there exist $\xi < \alpha$ and $\eta \in \kappa_i$ such that $F(\eta, \xi) = \gamma$. For such ξ we have $f(\xi + 1) \geq \gamma$. Hence $cf(\kappa) \leq \alpha < \kappa$. □

Corollary 18: *Every infinite successor cardinal is regular.*

Proof: This follows from Corollary 15 and Lemma 17. □

Corollary 19: $\aleph_\omega$ *is singular.*

Now it is also easy to see that such limit cardinals as $\aleph_{\omega\cdot\omega}, \aleph_{\omega_1}, \aleph_{\omega_4+\omega_3}, \aleph_{\omega_\omega}$ are all singular.

Does there exist an uncountable, regular limit cardinal? This is a tough question.

Definition 20: A cardinal κ is called *weakly inaccessible* if κ is an uncountable regular limit cardinal.

EXERCISE 27(PG): Show that if α is a weakly inaccessible cardinal, then $\alpha = \aleph_\alpha$.

EXERCISE 28(R): Show that the smallest ordinal α such that $\alpha = \aleph_\alpha$ is not a weakly inaccessible cardinal.

Were it not for the specific requirement of uncountability, then ω would be weakly inaccessible. Weakly inaccessible cardinals are our first examples of so-called *large cardinals.* Large cardinals are usually defined by taking a property of ω and generalizing it to uncountable cardinals. By all reasonable criteria, ω is much, much larger than all finite cardinals. Similarly, large cardinals turn out to be much, much larger than all preceding cardinals. A weakly inacessible cardinal κ for example is neither the successor of a smaller cardinal, nor the union of a set of fewer than κ smaller cardinals.

Large cardinals form a hierarchy. In this text, you will also encounter weakly compact cardinals and measurable cardinals. The first weakly compact cardinal is weakly inaccessible and much larger than the first weakly inaccessible cardinal. The first measurable cardinal is weakly compact and much larger than the first weakly compact cardinal. And so on.

It looks like we could go on forever and define larger and larger cardinals, but there is a catch. We shall show in the next chapter that it is not possible to prove in ZFC that even weakly inaccessible cardinals exist. Moreover, we shall show that the consistency of ZFC is provable in the theory ZFC + "There exists a weakly inaccessible cardinal." The latter has a rather disturbing consequence.

Fact 21: *If* ZFC + CON(ZFC) *is a consistent theory, then it is not the case that the existence of a weakly inaccessible cardinal is consistent relative to* ZFC.

Proof: Assume by way of contradiction that ZFC $\vdash$ CON(ZFC) $\to$ CON(ZFC + "There exists a weakly inaccessible cardinal.") Then ZFC + CON(ZFC) $\vdash$ CON(ZFC + "There exists a weakly inaccessible cardinal"), and since CON(ZFC) is provable in the theory ZFC + "There exists a weakly inaccessible cardinal," we also have ZFC + CON(ZFC) $\vdash$ CON(ZFC + CON(ZFC)). By Gödel's Second Incompleteness Theorem, the latter implies that ZFC + CON(ZFC) is an inconsistent theory, contradicting one of our assumptions. □

Note that Fact 21 does *not* imply that the existence of weakly inaccessible cardinals is inconsistent with ZFC or ZFC + CON(ZFC). It only implies that if we assume that the existence of a weakly inaccessible cardinal is consistent, then the gentlemen's agreement of Chapter 6 forces us to be somewhat suspicious of results obtained under such an assumption. In particular, if a theorem has been proved under the assumption that the existence of a weakly inaccessible cardinal is consistent with ZFC, then this assumption has to be stated explicitly in the text of the theorem. Similar remarks apply to other large cardinals. The consistency of the theory ZFC + "There exists a weakly inaccessible cardinal" can be proved in the theory ZFC + "There exists a weakly compact cardinal," and consistency of the latter theory is provable in ZFC + "There exists a measurable cardinal." The larger a cardinal one uses, the more suspicious one should be about the results proved with its help.

Let $\{A_i : i \in I\}$ be an indexed family of sets, and for all $i \in I$, denote $|A_i|$ by κ_i. Then the cardinality of the Cartesian product $\prod_{i\in I} A_i$ is denoted by $\prod_{i\in I} \kappa_i$. Unfortunately, this notation is ambiguous (think about the case where $A_i = \kappa_i$ for all $i \in I$), and one often needs to infer from the context what is meant. The proof of the next theorem provides ample opportunities to practice such context-sensitivity.

Theorem 22 (König): *Let* $(\kappa_i)_{i<\alpha}$ *and* $(\lambda_i)_{i<\alpha}$ *be sequences of cardinals such that* $\kappa_i < \lambda_i$ *for all* $i < \alpha$. *Then*

$$\sum_{i<\alpha} \kappa_i < \prod_{i<\alpha} \lambda_i.$$

Proof: Let $\{A_i : i < \alpha\}$ be a family of pairwise disjoint sets such that $|A_i| = \kappa_i$ for all $i < \alpha$, and let $A = \bigcup_{i<\alpha} A_i$. For $i < \alpha$, let π_i be a bijection from A_i onto

$\lambda_i \setminus \{0\}$. We define a function $F : A \to \prod_{i<\alpha} \lambda_i$ by

$$F(a)(i) = \begin{cases} \pi_i(a) & \text{if } a \in A_i; \\ 0 & \text{otherwise.} \end{cases}$$

EXERCISE 29(G): Convince yourself that F is one-to-one, and conclude that $\sum_{i<\alpha} \kappa_i \leq \prod_{i<\alpha} \lambda_i$.

Now suppose toward a contradiction that $\sum_{i<\alpha} \kappa_i = \prod_{i<\alpha} \lambda_i$; and let $\pi : A \to \prod_{i<\alpha} \lambda_i$ be a bijection. For $i < \alpha$ let $B_i = \{\pi(a)(i) : a \in A_i\}$. Since $|B_i| \leq \kappa_i$, for each $i < \alpha$ one can pick a $\gamma_i \in \lambda_i \setminus B_i$. But then $(\gamma_i)_{i<\alpha} \in \prod_{i<\alpha} \lambda_i \setminus rng(\pi)$, contradicting our assumption on π. □

EXERCISE 30(G): Convince yourself that in Theorem 22 it is not possible to replace the assumption that $\kappa_i < \lambda_i$ for all $i < \alpha$ by the weaker assumption that $\kappa_i \leq \lambda_i$ for all $i < \alpha$, and $\kappa_i < \lambda_i$ for at least one i, even if all κ_i's are assumed infinite.

Let us consider two special cases of König's Theorem. Let α be a cardinal, and let $\kappa_i = 1$ and $\lambda_i = 2$ for all $i < \alpha$. It is not hard to see that $\sum_{i<\alpha} 1 = \alpha$ and $\prod_{i<\alpha} 2 = 2^\alpha$. Since $|\mathcal{P}(\alpha)| = 2^\alpha$, Theorem 3.5 is an easy consequence of Theorem 22.

Now let λ be an infinite cardinal, and let $\alpha = cf(\lambda)$. Fix an increasing function $f : \alpha \to \lambda$ whose range is cofinal in λ, and let $\kappa_i = |f(i) \setminus \bigcup_{j<i} f(j)|$. Then $\kappa_i < \lambda$ for all $i < \alpha$, and König's Theorem yields the inequality

$$\sum_{i<\alpha} \kappa_i < \prod_{i<\alpha} \lambda.$$

Since the left hand side of the above inequality is equal to λ and the right hand side is equal to $\lambda^{cf(\lambda)}$, we have proved the following:

Theorem 23: *For every infinite cardinal λ, the inequality $\lambda < \lambda^{cf(\lambda)}$ holds.*

Corollary 24: *$cf(2^\kappa) > \kappa$ for every infinite cardinal κ.*

Proof: Let κ be an infinite cardinal. If $cf(2^\kappa) \leq \kappa$, then $(2^\kappa)^{cf(2^\kappa)} \leq (2^\kappa)^\kappa = 2^{\kappa \cdot \kappa} = 2^\kappa$, but the latter contradicts Theorem 23. □

What else can be proved about the operation $\kappa \mapsto 2^\kappa$? So far, we have established that

(i) $\forall \kappa, \lambda \, (\kappa \leq \lambda \to 2^\kappa \leq 2^\lambda)$ (Theorem 3.13(f3));
(ii) $\forall \kappa \geq \aleph_0 \; cf(2^\kappa) > \kappa$ (Corollary 24).

It turns out that this is all that can be proved about the operation $\kappa \mapsto 2^\kappa$ for regular infinite cardinals κ.

Theorem 25 (Easton): *Let* **F** *be a functional class such that the domain of* **F** *is the class of all regular infinite cardinals, the range of* **F** *is contained in the class of infinite cardinals, and* **F** *satisfies the following conditions for all κ, λ in its domain:*

(i) $\kappa \leq \lambda \to \mathbf{F}(\kappa) \leq \mathbf{F}(\lambda)$;

(ii) $cf(\mathbf{F}(\kappa)) > \kappa$.

Then it is relatively consistent with ZFC *that* $2^\kappa = \mathbf{F}(\kappa)$ *for all infinite regular cardinals* κ.[1]

Let us see which patterns of cardinal exponentiation Easton's Theorem allows. We mentioned already in Chapter 6 that the Continuum Hypothesis is relatively consistent with ZFC.

EXERCISE 31(G): Convince yourself that the Continuum Hypothesis as introduced in Chapter 3 is equivalent to the equality $2^{\aleph_0} = \aleph_0^+$.

Our formulation of Easton's Theorem implies in particular that one can consistently assume that $2^\kappa = \kappa^+$ for all infinite regular cardinals κ. In the next section we shall study the statement

(GCH) $\qquad 2^\kappa = \kappa^+$ for every infinite cardinal κ,

which is also relatively consistent with ZFC.[2] The letters GCH stand for "Generalized Continuum Hypothesis."

Easton's Theorem tells us that cardinal exponentiation can behave in much more bizarre ways. For example, it is perfectly possible that $2^{\aleph_0} = 2^{\aleph_1} = \cdots = 2^{\aleph_{17}} = \aleph_{19}$ and $2^{\aleph_{18}} = \aleph_{\omega+5}$. Or, it could be the case that $2^{\aleph_n} = \aleph_{n+1}$ for $n < 23$, and $2^{\aleph_{23}} = \aleph_{144}$. In the latter case we would say that $\aleph_{23}$ *is the first cardinal where* GCH *fails.*

EXERCISE 32(G): Which of the following patterns are consistent with ZFC?

(a) $2^{\aleph_0} = 2^{\aleph_1} = 2^{\aleph_2} = 2^{\aleph_3} = \aleph_2$;

(b) $2^{\aleph_0} = 2^{\aleph_1} = \aleph_5$ and $2^{\aleph_2} = 2^{\aleph_3} = \aleph_\omega$;

(c) $2^{\aleph_0} = \aleph_1$, $2^{\aleph_1} = \aleph_3$, $2^{\aleph_2} = 2^{\aleph_3} = \aleph_{\omega_1+16}$.

Easton's Theorem does not tell us much about 2^κ when κ is singular. In particular, suppose that $2^\kappa = \kappa^+$ for all regular infinite κ. Which values can $2^{\aleph_\omega}$ take? By (i), $2^{\aleph_\omega} \leq 2^{\aleph_{\omega+1}} = \aleph_{\omega+2}$. By (ii), $2^{\aleph_\omega} \geq \aleph_\omega^+ = \aleph_{\omega+1}$, but which of the two values is it? Are both possibilities consistent with ZFC? Can a singular cardinal be the first where GCH fails? Some highlights of the extensive research sparked by these problems will be discussed in the Mathographical Remarks at the end of this chapter. For now, let us just remark that in the models constructed in the original proof of Easton's Theorem, if κ is singular then 2^κ always takes the smallest value consistent with conditions (i), (ii), and the values of 2^λ for regular λ.

[1]The formulation of Easton's Theorem given here is visually appealing but slightly imprecise. A completely rigorous formulation of the theorem requires familiarity with the forcing method, which is beyond the scope of this book. The theorem says that if φ is a formula of two variables such that ZFC proves that the class $\mathbf{F} = \{\langle x, y\rangle : \varphi(x, y)\}$ satisfies all the assumptions of the theorem, then it is relatively consistent with ZFC that $\varphi(\kappa, \lambda) \to 2^\kappa = \lambda$ for all κ, λ. In order for this to work, one must also require that φ does not contain any reference to cardinal exponentiation. E.g., the formula "x is a regular cardinal and $y = x^{++}$" will work, but the formula "x is a regular cardinal and $y = (2^x)^+$" will not. While the property of "not containing a reference to cardinal exponentiation" is still a bit vague, it will suffice for most practical purposes.

[2]In fact, the relative consistency of GCH with ZFC was proved more than two decades earlier than Easton's Theorem.

EXERCISE 33(G): Suppose $2^{\aleph_n} = \aleph_{\omega+3n}$ for every $n \in \omega$. What is the smallest possible value of $2^{\aleph_\omega}$?

We conclude this chapter with a discussion of powers κ^λ for $\kappa > 2$.

Theorem 26: *Let λ be an infinite cardinal and assume that $2 \leq \kappa \leq \lambda$. Then $\kappa^\lambda = 2^\lambda$.*

Proof: By Theorems 11 and 3.13, the following inequalities hold: $2^\lambda \leq \kappa^\lambda \leq \lambda^\lambda \leq (2^\lambda)^\lambda = 2^{\lambda\cdot\lambda} = 2^\lambda$. □

Theorem 27 (Hausdorff): *Let κ, λ be infinite cardinals. Then $(\kappa^+)^\lambda = \kappa^\lambda \cdot \kappa^+$.*

Proof: If $\kappa^+ \leq \lambda$, then by Theorem 26, $\kappa^\lambda = (\kappa^+)^\lambda$, and by Theorem 11, $\kappa^\lambda = \kappa^\lambda \cdot \kappa^+$.

If $\lambda \leq \kappa$, then by regularity of κ^+, each function $f \in {}^\lambda(\kappa^+)$ is bounded in κ^+, and thus

$$|{}^\lambda(\kappa^+)| = |\bigcup_{\alpha<\kappa^+} {}^\lambda\alpha| \leq \sum_{\alpha<\kappa^+} |{}^\lambda\alpha| \leq \sum_{\alpha<\kappa^+} \kappa^\lambda = \kappa^\lambda \cdot \kappa^+.$$

By the use of α (instead of λ, μ, or ν) for the summation index we indicate that the sum is taken over all ordinals $< \kappa$, not just cardinals. □

As an application of Theorem 27, let us try to determine the value of $\aleph_4^{\aleph_2}$ under the assumption that GCH holds. Using Theorem 27 twice, we get: $\aleph_4^{\aleph_2} = \aleph_3^{\aleph_2} \cdot \aleph_4 = \aleph_2^{\aleph_2} \cdot \aleph_3 \cdot \aleph_4$. Now the combination of Theorem 26, GCH, and Theorem 11 implies that $\aleph_2^{\aleph_2} \cdot \aleph_3 \cdot \aleph_4 = 2^{\aleph_2} \cdot \aleph_3 \cdot \aleph_4 = \aleph_3 \cdot \aleph_3 \cdot \aleph_4 = \aleph_4$.

EXERCISE 34(G): Assume that $2^{\aleph_n} = \aleph_{2n}$ for all $n \in \omega$. Find $\aleph_7^{\aleph_3}$ and $\aleph_7^{\aleph_4}$.

Theorem 28 (Tarski): *If κ is a limit cardinal and $0 < \lambda < cf(\kappa)$, then $\kappa^\lambda = \sum_{\alpha<\kappa} |\alpha|^\lambda$.*

EXERCISE 35(PG): Prove Theorem 28.

EXERCISE 36(PG): Suppose GCH holds. Compute $\aleph_{\omega_1}^{\aleph_0}$.

Theorem 29 (Bukovský-Hechler): *Let κ, λ be infinite cardinals such that $cf(\kappa) \leq \lambda$ and $cf(\kappa) < \kappa$. Denote $\sum_{\alpha<\kappa} |\alpha|^\lambda = \mu$.*

(a) *If there exists $\alpha_0 < \kappa$ such that $|\alpha|^\lambda = |\alpha_0|^\lambda$ for all α satisfying $\alpha_0 \leq \alpha < \kappa$, then $\kappa^\lambda = \mu$.*

(b) *If for each $\alpha < \kappa$ there exists β such that $\alpha < \beta < \kappa$ and $\alpha^\lambda < \beta^\lambda$, then $\kappa^\lambda = \mu^{cf(\mu)}$.*

Theorems 26–29 allow us to calculate the values of κ^λ for all infinite cardinals κ, λ as long as we know the value of $\mu^{cf(\mu)}$ for every infinite cardinal μ.

EXERCISE 37(G): Assume that $\mu^{cf(\mu)} = \mu^+$ for every infinite cardinal μ. Find $\aleph_{\omega_1}^{\aleph_3}$.

Proof of Theorem 29: (a) Let α_0 be as in the assumption of this case. Then $|\alpha_0|^\lambda \geq \kappa$.

EXERCISE 38(G): Why?

Therefore, $\mu = \sum_{\alpha<\kappa} |\alpha|^\lambda \leq |\alpha_0|^\lambda \cdot \kappa = |\alpha_0|^\lambda \leq \mu$. Hence, $\mu = |\alpha_0|^\lambda \leq \kappa^\lambda \leq (|\alpha_0|^\lambda)^\lambda = |\alpha_0|^{\lambda\cdot\lambda} = |\alpha_0^\lambda| = \mu$.

(b) Note that in this case $cf(\mu) = cf(\sum_{\alpha<\kappa} |\alpha|^\lambda) = cf(\kappa)$. Moreover, $\mu^{cf(\kappa)} = (\sum_{\alpha<\kappa} |\alpha|^\lambda)^{cf(\kappa)} \leq (\kappa^\lambda \cdot \kappa)^{cf(\kappa)} = (\kappa^\lambda)^{cf(\kappa)} = \kappa^{\lambda\cdot cf(\kappa)} = \kappa^\lambda$. Thus our task boils down to showing that $\kappa^\lambda \leq \mu^{cf(\kappa)}$. For this, denote by F the family of all functions from λ into some ordinal $\alpha < \kappa$. Clearly, $|F| \leq \mu$. We are going to construct an injection $H : {}^\lambda\kappa \rightarrow {}^{cf(\kappa)}F$. We choose an increasing sequence $\langle \gamma_\xi : \xi < cf(\kappa)\rangle$ of ordinals such that $\lim_{\xi\rightarrow cf(\kappa)} \gamma_\xi = \kappa$. For $f \in {}^\lambda\kappa$ and $\xi < cf(\kappa)$, we let $H(f)(\xi)$ be the function $g \in F$ such that

$$g(\beta) = \begin{cases} f(\beta) & \text{if } f(\beta) < \gamma_\xi, \\ 0 & \text{otherwise.} \end{cases}$$

EXERCISE 39(G): Convince yourself that H is one-to-one and maps ${}^\lambda\kappa$ into ${}^{cf(\kappa)}F$. Conclude that $\kappa^\lambda \leq \mu^{cf(\kappa)}$. □

Mathographical Remarks

Much of the research done on cardinal arithmetic has been inspired by the famous *Singular Cardinal Hypothesis*, which is usually abbreviated as SCH. This hypothesis states that $\mu^{cf(\mu)} = \max\{2^{cf(\mu)}, \mu^+\}$ whenever μ is a singular cardinal. If SCH holds, then for every singular cardinal μ the power 2^μ is the smallest value consistent with conditions (i) and (ii) of Easton's Theorem and with the behavior of the operation $\kappa \mapsto 2^\kappa$ on regular cardinals κ.

In the early 1970's, Prikry and Silver showed that if the existence of certain large cardinals is consistent with ZFC, then so is the negation of SCH (see K. Prikry, *Changing measurable into accessible cardinals,* Diss. Math. 68 (1970), 5–52, and J. Silver, *Reversed Easton iteration and failure of GCH at supercompacts,* unpublished notes). In 1977, M. Magidor showed in his paper *On the singular cardinals problem I,* Israel J. Math. 28 (1) (1977), 1–31, that the SCH may fail at $\aleph_\omega$. His construction also uses an assumption that certain large cardinals exist. By a result of Jensen (see K. J. Devlin and R. B. Jensen, *Marginalia to a theorem of Silver,* ISILC Logic Conf. Kiel 1974, Lecture Notes in Math. 499, Springer, 1975, 115–142), the use of large cardinals in the construction of models for the failure of SCH is necessary.

It is possible to obtain in ZFC some upper bounds for $\mu^{cf(\mu)}$ for singular cardinals μ. At the time of this writing, the best known upper bound for $\aleph_\omega^{\aleph_0}$ is $\max(2^{\aleph_0}, \aleph_{\aleph_4})$. It was obtained by S. Shelah in 1989. For expositions of this result, see T. J. Jech, *Singular Cardinal Problem: Shelah's Theorem on $2^{\aleph_\omega}$*, Bull. London Math. Soc. 24 (1992), 127–139, or M. R. Burke and M. Magidor, *Shelah's pcf Theory and its Applications,* Ann. Pure and Appl. Logic 50 (1990), 207–254.

Another central problem used to be the question whether the first cardinal where GCH fails can be singular. The models of Prikry and Silver and of Magidor's paper mentioned above violate GCH even below the first cardinal μ where SCH fails. In 1975, Silver showed in his paper *On the singular cardinals problem,* Proc. of the International Congress of Math. Vancouver, vol. 1 (1974), 265–268, that the first cardinal where GCH fails cannot be a singular cardinal of uncountable cofinality; and Magidor proved in his paper *On the singular cardinal problem II,* Ann. of Math. 106 (1977), 517–647, that if the existence of certain large cardinals is consistent with ZFC, then it is possible that $\aleph_\omega$ is the first cardinal where GCH fails. For a more detailed exposition of cardinal arithmetic we recommend T. J. Jech's book *Set Theory,* Academic Press 1978, and, at an advanced level, S. Shelah's *Cardinal arithmetic,* Oxford University Press, 1994.

CHAPTER 12

Pictures of the Universe

Throughout this chapter, we will work in the theory ZF.

Imagine you are the architect of the universe of sets. Your task is to build up this universe step by step. At each step of the construction, you may build new sets only from elements that have already been constructed at previous steps. How could you proceed?

Let us denote the part of the universe built until step α by V_α. In the beginning, nothing exists, that is, $V_0 = \emptyset$. How can you get started? The only set that does not require preexisting building blocks is the empty set. So, after the first step, your universe will look like this: $V_1 = \{\emptyset\}$. In the next step, you have already one building block, and this allows you to form the set $\{\emptyset\}$. So you get: $V_2 = \{\emptyset, \{\emptyset\}\}$. Next you can form the sets $\{\{\emptyset\}\}$ and $\{\emptyset, \{\emptyset\}\}$, which yields: $V_3 = \{\emptyset, \{\emptyset\}, \{\{\emptyset\}\}, \{\emptyset, \{\emptyset\}\}\}$.

EXERCISE 1(G): List all sets that you are allowed to form in the fourth step, and write out the corresponding V_4.

Note that $V_{n+1} = \mathcal{P}(V_n)$ for $n < 4$. So why not continue like this and define recursively $V_{n+1} = \mathcal{P}(V_n)$ for all $n \in \omega$? This makes sense, since the only sets you are allowed to add at stage $n+1$ are sets composed of elements of V_n, that is, subsets of V_n. Moreover, you get an increasing sequence (see Theorem 2(c)).

Proceeding like this, after ω steps you will have built a collection $V_\omega = \bigcup_{n\in\omega} V_n$. We used the symbol V_ω already in Chapter 7, where we mentioned that $\langle V_\omega, \overline{\in}\rangle$ is a standard model for all the axioms of ZFC, except for the Axiom of Infinity (see Exercise 7.32). Since you want the Axiom of Infinity to hold in the universe $\mathbf{V}$ of all sets, you cannot stop at this point.

In the next step of the construction, you are allowed to add all subsets of V_ω to the universe, then all subsets of subsets of V_ω, and so on. After ω more steps, you will have constructed a partial universe $V_{\omega+\omega}$. Unfortunately, this is still not a model of ZFC (see Exercise 24). By now you probably realize that the "steps" involved in the construction of $\mathbf{V}$ are the ordinals, and the "set-theoretic timeline" mentioned in the Introduction is the class **ON**. You have a long way to go before you can relax and contemplate the whole universe $\mathbf{V}$ ... .

The construction itself can be described in a remarkably simple way:

Definition 1: Define by recursion over the class **ON**:
$V_0 = \emptyset$;
$V_{\alpha+1} = \mathcal{P}(V_\alpha)$;
$V_\alpha = \bigcup_{\beta<\alpha} V_\beta$ for $\alpha \in \mathbf{LIM}$.

Note that if V_α is a set, then so is $V_{\alpha+1}$ (by the Power Set Axiom); and if $\alpha > 0$ is a limit ordinal, then the collection $\{V_\beta : \beta < \alpha\}$ is a set by the Replacement Axiom Schema (as long as V_β is a set for each $\beta < \alpha$), and the Union Axiom implies that V_α is also a set. Since $V_0 = \emptyset$ is a set, induction over the class **ON** shows that each V_α is a set. Thus the recursion in Definition 1 defines a functional class $\{\langle \alpha, V_\alpha \rangle : \alpha \in \mathbf{ON}\}$. This functional class is called the *cumulative hierarchy*. The next theorem summarizes the basic properties of this hierarchy. Before stating it, let us refresh our memories regarding the notion of a transitive set.

EXERCISE 2(G): Recall that a set z is transitive if $TC(z) = z$ (Definition 4.4(c)). Show that a set z is transitive if and only if $x \subseteq z$ whenever $x \in z$.

Theorem 2: *Let $\alpha \in \mathbf{ON}$. Then:*
(a) *For every set x, $x \subseteq V_\alpha$ iff $x \in V_{\alpha+1}$.*
(b) *V_α is a transitive set.*
(c) *$V_\beta \subset V_\alpha$ for every $\beta < \alpha$.*
(d) *If $y \subseteq x$ and $x \in V_\alpha$, then $y \in V_\alpha$.*

Proof: Point (a) follows immediately from Definition 1. Let us prove (b) by induction over α. Clearly, $V_0 = \emptyset$ is a transitive set. Now suppose $\alpha > 0$ and V_β is transitive for all $\beta < \alpha$. Let $x \in V_\alpha$. We want to show that $x \subseteq V_\alpha$. If α is a limit ordinal, then $x \in V_\beta$ for some $\beta < \alpha$, and the inductive assumption implies that $x \subseteq V_\beta \subseteq V_\alpha$. If $\alpha = \beta + 1$ for some β, then $x \subseteq V_\beta$ by (a), and transitivity of V_β implies that every element y of x is an element of V_β as well. Invoking transitivity of V_β again, we see that every element y of x is also a subset of V_β. Now (a) implies that every element of x is also an element of V_α, which is another way of saying that $x \subseteq V_\alpha$. □

EXERCISE 3(G): Prove points (c) and (d) of Theorem 2.

The cumulative hierarchy neatly formalizes the idea of constructing sets step by step. But does *every* set appear at some step of the hierarchy?

Theorem 3: *For every set x there exists an $\alpha \in \mathbf{ON}$ such that $x \in V_\alpha$.*

Proof: Suppose not, and let $x \in \mathbf{V} \backslash \bigcup_{\alpha \in \mathbf{ON}} V_\alpha$. Since $x \subseteq TC(x)$, it follows from Theorem 2(d) that $TC(x) \notin \bigcup_{\alpha \in \mathbf{ON}} V_\alpha$. So, without loss of generality, we may assume that x is transitive. An instance of the Comprehension Axiom Schema entitles us to form the set $x_0 = \{y \in x : \exists \alpha \in \mathbf{ON}\, y \in V_\alpha\}$. Define a function $\varrho : x_0 \to \mathbf{ON}$ by $\varrho(y) = \min\{\alpha : y \in V_\alpha\}$. By an instance of the Replacement Axiom Schema, $rng(\varrho)$ is a set of ordinals. Let $\delta = \bigcup rng(\varrho)$. Then $\delta \in \mathbf{ON}$ (by Theorem 10.10), and $x_0 \subseteq V_{\delta+1}$. By Theorem 2(a), $x_0 \in V_{\delta+2}$, and it follows that $x \backslash x_0 \neq \emptyset$. By the Foundation Axiom, there exists a $y_0 \in x \backslash x_0$ such that $y_0 \cap x \backslash x_0 = \emptyset$. By transitivity of x, this implies that $y_0 \subseteq x_0$. But then Theorem 2(d) implies $y_0 \in V_{\delta+2}$, hence $y_0 \in x_0$, contradicting the choice of y_0. □

Theorem 3 can be restated as follows: $V = \bigcup_{\alpha \in \mathbf{ON}} V_\alpha$.

But isn't this self-evident? We started this chapter by asking you to play the role of the architect of the universe, and we explained why it is natural for you to build up the cumulative hierarchy. Now, if you call this process "building the universe **V**," then, of course, **V** must turn out to be equal to $\bigcup_{\alpha \in \mathbf{ON}} V_\alpha$.

And isn't this whole construction process circular anyway? An ordinal α gets constructed only at stage α (see Corollary 6 below). But as the architect of the universe, you must know beforehand that there is such a thing as a step α before you can construct anything at this step. In other words, the class of ordinals would have to be given before you start your construction.

What is wrong with the "architect's view" of set theory? Does it perhaps lead to a contradiction similar to Russell's Paradox? As far as we know, nothing is wrong with it. The tale of the architect of the universe is only a metaphor. Like any other metaphor, it can only explain some aspects of reality. You can push this metaphor a little further by imagining that in the beginning the architect is immersed in a kind of primordial soup of sets and strives to arrange this universe in a neat hierarchy. Looked at in this way, Theorem 3 says that the hierarchy built according to Definition 1 comprises the whole universe.[1] But that is about as far as the metaphor of the architect of the universe can take us. It can give us some idea why the axioms of ZFC look the way they do, but it cannot replace the axioms as a kind of First Principle to which ZFC could be reduced.

Now let us leave the lofty realm of ontology and show how the cumulative hierarchy can help in studying sets. Theorem 3 allows us to define a rank function on the class of all sets.

Definition 4: For each set x let $rank(x) = \min\{\alpha : x \in V_{\alpha+1}\}$.

Let **RANK** denote the functional class $\{\langle x, rank(x)\rangle : x \in \mathbf{V}\}$.

Claim 5: *For every transitive set z, the restriction of* **RANK** *to z is the rank function of the wellfounded relation $\in \cap\, z \times z$.*

EXERCISE 4(PG): Prove Claim 5.

Corollary 6: *$rank(\alpha) = \alpha$ for each ordinal α.*

Proof: Apply Claim 5 to $z = \alpha + 1$ and note that the rank function on an ordinal is the identity. Corollary 6 can also be proved directly by induction over α. □

Corollary 7: *Let z be a transitive set. If $rank(z) = \alpha$, then for every $\beta < \alpha$ there exists an $x \in z$ such that $rank(x) = \beta$.*

Proof: This follows immediately from Exercise 10.15(b). □

Now we are going to present several important applications of the cumulative hierarchy. Our first application is a characterization of proper classes.

Claim 8: *Let **X** be a class. Then **X** is a set iff $\mathbf{X} \subseteq V_\alpha$ for some α.*

[1] In versions of set theory where the Foundation Axiom is not assumed, some dissident sets may remain outside the cumulative hierarchy.

Proof: If $\mathbf{X} \subseteq V_\alpha$ for some α, then an instance of the Separation Axiom Schema applied to a representation of $\mathbf{X}$ allows us to conclude that $\mathbf{X}$ is a set. On the other hand, if $\mathbf{X}$ is a set, then by Theorem 3, there exists an α such that $\mathbf{X} \in V_\alpha$, and hence, by Theorem 2, $\mathbf{X} \subseteq V_\alpha$. □

Definition 9: Let λ be a cardinal. By H_λ we denote the class of sets *hereditarily of cardinality less than* λ; i.e., we define $H_\lambda = \{x : |TC(x)| < \lambda\}$. In particular, $H_{\aleph_0}$ is the class of *hereditarily finite* sets and $H_{\aleph_1}$ is the class of *hereditarily countable* sets.[2]

It is not immediately clear from the above definition whether the H_λ's are sets or proper classes. The next theorem settles this matter.

Theorem 10: *Let λ be an infinite cardinal. Then $H_\lambda \subseteq V_\lambda$. In particular, each H_λ is a set.*

Proof: Fix λ, assume $x \in H_\lambda$, and let $z = TC(x)$. Then z is transitive and $|z| < \lambda$. It follows from Corollary 7 that $rank(z) < \lambda$, which implies that $z \in V_\lambda$. By Theorem 2(d), also $x \in V_\lambda$. Thus $H_\lambda \subseteq V_\lambda$, and by Claim 8, H_λ is a set. □

Corollary 11: $V_\omega = H_{\aleph_0}$.

Proof: It suffices to show that $V_\omega \subseteq H_{\aleph_0}$. To accomplish this, one can show by induction that each V_n is contained in $H_{\aleph_0}$.

EXERCISE 5(G): Show that if $x \in H_{\aleph_0}$, then $\mathcal{P}(x) \in H_{\aleph_0}$. □

Now let us reconsider the definition of a cardinal. The concept of a cardinal as an initial ordinal developed in Chapter 11 requires the Axiom of Choice. Of course, even if (AC) fails, there are initial ordinals, but not every set is equipotent with one of them. So how can we define cardinals without using (AC)? In Chapter 8 we associated with every set x a class $\mathbf{CARD}(x) = \{y : y \approx x\}$ that corresponds to the intuitive notion of a cardinal developed in Chapter 3. The definition of $\mathbf{CARD}(x)$ does not require the Axiom of Choice. Unfortunately, this notion fits awkwardly into the framework of axiomatic set theory, because for nonempty x the class $\mathbf{CARD}(x)$ is proper. The cumulative hierarchy allows us to define a notion of cardinals in ZF so that the cardinals are sets rather than proper classes.

Definition 12: For any set x define:

$$c(x) = \min\{\alpha : \exists y \in V_\alpha\,(y \approx x)\};$$
$$||x|| = \{y \in V_{c(x)} : y \approx x\}.$$

EXERCISE 6(G): (a) Show that $||x||$ is a set for every x.
(b) Show that $||x|| = ||y||$ iff $x \approx y$.

EXERCISE 7(G): Working in ZF alone, define a functional class **ORD** in such a way that

[2]Many authors use the notation $H(\lambda)$ instead of H_λ.

(i) the domain of **ORD** is the class **PO** of all p.o.'s;
(ii) $\forall x, y \in \mathbf{PO}\,(\exists z\,(\langle x, z\rangle, \langle y, z\rangle \in \mathbf{ORD}) \leftrightarrow x \cong y)$.

It is instructive to compare Exercise 7 with Exercise 8.16. Note that the class **OT** constructed in Exercise 8.16 has slightly stronger properties than the class **ORD** of Exercise 7, but was constructed with the help of a very strong version of the Axiom of Choice (compare Corollary 33 below).

EXERCISE 8(G): Prove that for every formula $\varphi(x, y)$ of L_S the following holds:[3]

$$(\mathbf{A11}\varphi) \qquad \forall a\,(\forall x \in a \exists y\, \varphi(x, y) \rightarrow \exists z \forall x \in a \exists y \in z\, \varphi(x, y)).$$

Now let us consider the following question: How could one prove theorems like "CON(ZF) $\rightarrow$ CON(ZFC)" or "CON(ZF) $\rightarrow$ CON(ZFC + GCH)?" In Chapter 6 we mentioned that one might start out with a model of ZF and modify it so as to obtain a model for ZFC or ZFC + GCH. It is also possible to start not with a model (whose universe should be a set) but with the universe of sets **V** itself (which is a proper class). Now suppose we have decided what kind of structure to start with. Then we have two possibilities: We can try to construct a nonstandard model $\langle \mathbf{X}, \mathbf{E}\rangle$ or a standard model $\langle \mathbf{X}, \overline{\in}\rangle$.[4] Up to date, nonstandard models of ZF have found only very few applications, and in this chapter we shall concentrate exclusively on standard models.

Since in a standard model the interpretation of the symbol $\in$ of L_S is always the "real" membership relation $\overline{\in}$, we shall often identify a standard model with its underlying class. Now we give a formal definition of standard models.

Definition 13: (a) Let **X** be a class, and let φ be a formula of L_S. The *relativization of* φ *to* **X** is the formula $\varphi_{/\mathbf{X}}$ that is obtained by replacing each quantifier "$\exists v_i$" occurring in φ by "$\exists v_i \in \mathbf{X}$," and each quantifier "$\forall v_i$" occurring in φ by "$\forall v_i \in \mathbf{X}$."

(b) If T is a theory in L_S and **X** is a class such that $\varphi_{/\mathbf{X}}$ holds for every φ in T, then we call **X** a *standard model of* T and write $\mathbf{X} \models T$.

The phrase "$\varphi_{/\mathbf{X}}$ holds" in the above definition is naively interpreted as "$\varphi_{/\mathbf{X}}$ is true in the universe **V**."

As we mentioned in Chapter 8, the use of symbols for proper classes is just a convenience, and can always be eliminated: Suppose $\mathbf{X} = \{x : \psi(x)\}$. Then the relativization $\varphi_{/\mathbf{X}}$ of a formula φ can be obtained by replacing all occurences of "$\exists v_i$" in φ by "$\exists v_i(\psi(v_i) \wedge \ldots)$, and all occurences of "$\forall v_i$" in φ by "$\forall v_i(\psi(v_i) \rightarrow \ldots)$."

[3]The formulas (A11φ) are instances of what is called the *Axiom Schema of Collection.* Note that the difference between Collection and Replacement is that in the former we do not require that there exists *exactly one* y such that $\varphi(x, y)$.

[4]The notions of standard and nonstandard model were introduced at the beginning of Chapter 7 for the case where **X** is a set. If **X** is a proper class, then, strictly speaking, we cannot form ordered pairs $\langle \mathbf{X}, y\rangle$. We use $\langle \mathbf{X}, \overline{\in}\rangle$ as a suggestive abbreviation, and we hope the reader will forgive us the notational transgression. A rigorous treatment of standard models whose universes are proper classes is given by Definition 13(b). In the remainder of this chapter, we will use the notation $\langle N, \overline{\in}\rangle$ only when we want to indicate that N is a set.

EXERCISE 9(R): Show that if $\mathbf{X}$ is a set, then the notion of $\mathbf{X} \models T$ established by Definition 13(b) coincides with the notion $\langle \mathbf{X}, \overline{\in} \rangle \models T$ defined in Chapter 5. *Hint:* We did something similar in the proof of Theorem 7.1.

EXERCISE 10(R): Define a notion of nonstandard models of L_S which are (possibly proper) classes. Make sure that if your models happen to be sets, then you get the satisfaction relation $\langle \mathbf{X}, \mathbf{E} \rangle \models T$ defined in Chapter 5.

If we get the same concept as in Chapter 5, why do we have a separate definition for class models? And why do we bother to look at class models in the first place?

One can prove in ZF that if a theory has a set model, then it is consistent. But by Gödel's Second Incompleteness Theorem it is impossible to prove in ZF that ZF is a consistent theory. Therefore, if we want to work with set models, then we have to transcend the theory ZF by making the unprovable assumption that there is such a model.

Class models are different in this respect. Let $\{\varphi_0, \dots, \varphi_n, \dots\}$ be a list of all axioms of ZF. Given $n \in \omega$, the symbols "$\mathbf{V} \models \{\varphi_0, \dots, \varphi_n\}$" and "$\varphi_0 \wedge \dots \wedge \varphi_n$" stand for exactly the same thing. Thus, for every $n \in \omega$, ZF $\vdash$ "$\mathbf{V} \models \{\varphi_0, \dots, \varphi_n\}$." Now common sense would dictate that $\mathbf{V} \models$ ZF, but the latter is *not* a theorem of ZF! How do we know? Well, the symbol "$\mathbf{V} \models$ ZF" would stand for something like "$\varphi_0 \wedge \dots \wedge \varphi_n \dots$," and this is not even a well-formed expression of L_S, let alone a theorem of ZF. In other words, it is not provable that $\mathbf{V}$ is a model of ZF, and no conflict with Gödel's Second Incompleteness Theorem arises.

So what is the meaning of Definition 13(b)? Shouldn't the use of "$\mathbf{X} \models T$" be outlawed when T is infinite? Maybe so. But consider the following situation: Suppose $\mathbf{X}$ is a class such that for all $n \in \omega$,

$$\text{ZF} \vdash (\text{AC})_{/\mathbf{X}} \wedge \text{GCH}_{/\mathbf{X}} \wedge \varphi_{0/\mathbf{X}} \wedge \dots \wedge \varphi_{n/\mathbf{X}}.$$

Moreover, it is conceivable that the following is a theorem of ZF:

(α) "For every $n \in \omega$, there exists a formal proof p of the statement $(\text{AC})_{/\mathbf{X}} \wedge \text{GCH}_{/\mathbf{X}} \wedge \varphi_{0/\mathbf{X}} \wedge \dots \wedge \varphi_{n/\mathbf{X}}$ such that the only axioms used in p are from the list $\{\varphi_n : n \in \omega\}$."

Note that such a theorem says nothing about the consistency of the list $\{\varphi_n : n \in \omega\}$ itself and hence does not contradict Gödel's Second Incompleteness Theorem. In order to prove a theorem like (α) on the purely syntactical level, one would have to show how to construct *formal* proofs of $\text{GCH}_{/\mathbf{X}}$, $\varphi_{0/\mathbf{X}}$, etc. straight from the axioms. But human minds are not designed to shuffle meaningless symbols. Fortunately, Definition 13(b) allows us to express (α) in the crisper form:

(β) "If $\mathbf{V} \models$ ZF then $\mathbf{X} \models$ ZFC + GCH."

It is much easier for us to understand what is going on if we imagine that we can start with a model $\mathbf{V}$ of ZF, and, by throwing away a few sets if need be, construct another class $\mathbf{X}$ that is a model of ZFC + GCH. In this sense, (β) is more user-friendly than (α). Once we have proved theorem (β) in this way, it is not hard to convince ourselves that (β) has a formal ZF-counterpart (α), and that it would be "in principle" possible to translate the proof of (β) into a ZF-proof of (α).

Most of the standard models considered in the set-theoretic literature are transitive. The reason is that transitive models are easier to work with than non-transitive ones. Moreover, every interesting consistency result that can be obtained by means of a standard model can also be obtained by means of a standard transitive model. More precisely, the following holds.

Theorem 14: *Let T be a theory in L_S that contains the Axiom of Extensionality. If T has a standard model, then T has a standard transitive model.*

Proof: We prove the case where T has a standard set model and leave the case of class models as an exercise. As in Chapter 7, we abbreviate the Axiom of Extensionality by (A1).

Claim 15: *Let N be a set. Then $\langle N, \overline{\in}\rangle \models$ (A1) iff $\overline{\in}$ is an extensional relation on N.*

EXERCISE 11(G): Recall Definition 4.38 (of an extensional relation) and prove Claim 15.

Now suppose T is a theory in L_S such that (A1) $\in T$, and assume that $\langle N, \overline{\in}\rangle \models T$. By Theorem 4.39, there exist a transitive set M (the Mostowski collapse of N) and a bijection $F : N \to M$ such that

(M) $$\forall x, y \in N\,(x \overline{\in} y \leftrightarrow F(x) \overline{\in} F(y)).$$

Condition (M) asserts that F is an isomorphism of the models $\langle N, \overline{\in}\rangle$ and $\langle M, \overline{\in}\rangle$. By Theorem 6.4, $\langle M, \overline{\in}\rangle \models T$.

All notions used in this proof can be generalized to classes. In particular, we call a proper class $\mathbf{X}$ *transitive* if $\forall x \in \mathbf{X} \forall y \in \mathbf{X}\,(y \in x \to y \in \mathbf{X})$.

EXERCISE 12(R): Prove that if T is as in the assumptions of Theorem 14, and if a class $\mathbf{X}$ is a standard model of T, then there exists a transitive class $\mathbf{Y}$ such that $\mathbf{Y} \models T$. □

Now suppose you are immersed in the universe of sets $\mathbf{V}$, you are being told that the axioms of ZF hold, and you want to find a model for ZFC or even ZFC + GCH. By what was said above, it would be wise to look for a standard, transitive model. Now you can look either for a set model or for a proper class. Let us first try the former. Earlier in this chapter we defined the V_α's and the H_λ's. These are transitive sets; are some of them perhaps models of ZF?

Exercise 7.32 settles this problem for $V_\omega = H_{\aleph_0}$.[5] If you did this exercise, you already know that it is rather tedious to verify all the axioms of ZF one by one. Fortunately, there are some shortcuts.

Lemma 16: *Let $\mathbf{X}$ be a transitive class. Then $\mathbf{X} \models$ (A1) $\wedge$ (A2) $\wedge$ (A3).*

Proof: Theorem 7.1 asserts that (A1) holds in all standard transitive (set) models; and by Exercise 7.6(b), the Empty Set Axiom holds in all standard (set) models.

[5] Recall that all axioms except the Axiom of Infinity hold in this model.

EXERCISE 13(PG): (a) Generalize the results of Theorem 7.1 and Exercise 7.6(b) to class models.

(b) Prove that $\mathbf{X} \models$ (A3) holds for every class $\mathbf{X}$. □

We say that a class $\mathbf{X}$ is *closed under the power set operation* if $\mathcal{P}(x) \in \mathbf{X}$ for every $x \in \mathbf{X}$.

EXERCISE 14(G): Convince yourself that V_α is closed under the power set operation if and only if $\alpha \in \mathbf{LIM}$.

Lemma 17: *Let N be a transitive set closed under the power set operation. Then $\langle N, \overline{\in}\rangle$ satisfies the Power Set Axiom and the Comprehension Axiom Schema.*

Proof: This may look utterly trivial, since the Power Set Axiom apparently "says" that N is closed under the power set operation. But not so fast. If $x, y \in N$, is it really true that

(E) $$y = \mathcal{P}(x) \quad \text{iff} \quad \langle N, \overline{\in}\rangle \models y = \mathcal{P}(x)\,?$$

Recall that expressions like "$\mathcal{P}(x)$" are abbreviations. To be on the safe side, let us unravel the meanings of both sides of (E). The expression "$y = \mathcal{P}(x)$" stands for:

(lhs) $$\forall z\,(z \subseteq x \leftrightarrow z \in y),$$

whereas "$\langle N, \overline{\in}\rangle \models y = \mathcal{P}(x)$" abbreviates:

(rhs) $$\forall z \in N\,(z \subseteq x \leftrightarrow z \in y).$$

There is no reason to expect that (lhs) should always be equivalent to (rhs). For example, if N is not transitive, there could be some z in y which is not a subset of x (and hence witnesses the failure of (lhs)), but is not an element of N either (and hence does not invalidate (rhs)). But wait a minute. What do we mean by "z is not a subset of x?" Do we want to interpret this as $z \not\subseteq x$ or as $\langle N, \overline{\in}\rangle \models z \not\subseteq x$? To get a handle on this confusing issue, we need a new concept.

Definition 18: Let $\varphi(v_0, \dots, v_k)$ be a formula of L_S with free variables $v_0, \dots, v_k$, and let $\mathbf{X}$ be a class. We say that φ is *absolute for* $\mathbf{X}$ if $\varphi(x_0, \dots, x_k) \leftrightarrow \varphi_{/\mathbf{X}}(x_0, \dots, x_k)$ for all $x_0, \dots, x_k \in \mathbf{X}$. We say that φ is *absolute for transitive models* if φ is absolute for every transitive class $\mathbf{X}$. A formula φ is *absolute* if it is absolute for every class $\mathbf{X}$.

Let us list some formulas whose absoluteness for transitive models will be used in this chapter.

Claim 19: *The following formulas are all absolute for transitive models:*

(a) $\varphi_0(x, y)$: $y = TC(x)$;
(b) $\varphi_1(x)$: $x \in \mathbf{ON}$;
(c) $\varphi_2(x, y)$: $x \subseteq y$;
(d) $\varphi_3(x, y)$: $y = \bigcup x$;
(e) $\varphi_4(x, y, z)$: $z = \{x, y\}$;
(f) $\varphi_5(x)$: $x = \omega$;
(g) $\varphi_6(x)$: $x \in \omega$;
(h) $\varphi_7(x, y, z)$: $x, y, z \in \mathbf{ON} \wedge x + y = z$;
(i) $\varphi_8(x)$: *x is a function;*
(j) $\varphi_9(x, y, z)$: *z is a bijection from x onto y.*

It is worth noting that none of the above formulas is absolute. This is perhaps most surprising for φ_4; so let us demonstrate nonabsoluteness of this formula. Consider the model $\langle N, \overline{\in}\rangle$, where $N = \{\emptyset, \{\emptyset\}, \{\emptyset, \{\emptyset\}, \{\{\emptyset\}\}\}\}$. Of course,

$$\{\emptyset, \{\emptyset\}, \{\{\emptyset\}\}\} \neq \{\emptyset, \{\emptyset\}\},$$

but nevertheless

$$\langle N, \overline{\in}\rangle \models \varphi_4(\emptyset, \{\emptyset\}, \{\emptyset, \{\emptyset\}, \{\{\emptyset\}\}\})$$

EXERCISE 15(PG): Show that none of the formulas φ_0–φ_9 is absolute.

Proof of Claim 19: We give the proof only for φ_2 and leave the other formulas as an exercise. Suppose $x, y \in N$ and N is transitive. Then the formula "$x \subseteq y$" is an abbreviation for

(A) $$\forall z\,(z \in x \to z \in y),$$

and "$\langle N, \overline{\in}\rangle \models x \subseteq y$" stands for

(B) $$\forall z \in N\,(z \in x \to z \in y).$$

It is clear that (A) implies (B), even if N is not transitive. But if N is transitive, then $z \in x$ already implies that $z \in N$. In other words, the implication $z \in x \to z \in y$ is vacuously true for all $z \notin N$ (because the antecedent fails), so that the truth or falsity of (A) only depends on the z's in N, which is another way of saying that (B) implies (A).

EXERCISE 16(PG): Prove the remaining parts of Claim 19.

Note that the formula "$y = \mathcal{P}(x)$" does not appear in Claim 19. For good reason: It is not absolute for transitive models.

EXERCISE 17(G): Let $\alpha \in \mathbf{ON}$. Show that $\mathbf{ON} \models \mathcal{P}(\alpha) = \alpha + 1$.

Luckily, the formula $y = \mathcal{P}(x)$ is absolute for transitive models closed under the power set operation.

EXERCISE 18(G): (a) Assume that N is transitive, $x, y \in N$, and $y = \mathcal{P}(x)$. Show that $\langle N, \overline{\in}\rangle \models$ "$y = \mathcal{P}(x)$."

(b) Deduce Lemma 17. □

EXERCISE 19(PG): Show that for every ordinal α and every infinite cardinal λ the models $\langle V_\alpha, \overline{\in}\rangle$ and $\langle H_\lambda, \overline{\in}\rangle$ satisfy the Union Axiom.

EXERCISE 20(G): Show that if $\alpha > \omega$ and λ is an uncountable cardinal, then $\langle V_\alpha, \overline{\in}\rangle$ and $\langle H_\lambda, \overline{\in}\rangle$ satisfy the Axiom of Infinity.

EXERCISE 21(G): Show that if α is a limit ordinal and λ is an infinite cardinal, then $\langle V_\alpha, \overline{\in}\rangle$ and $\langle H_\lambda, \overline{\in}\rangle$ satisfy the Pairing Axiom.

EXERCISE 22(G): Let $\alpha \in \mathbf{ON}$, and let λ be an infinite cardinal. Show that if (AC) holds in $\mathbf{V}$, then $\langle V_\alpha, \overline{\in}\rangle$ and $\langle H_\lambda, \overline{\in}\rangle$ satisfy the Axiom of Choice.

EXERCISE 23(G): Show that if λ is a regular infinite cardinal, then $\langle H_\lambda, \overline{\in}\rangle$ satisfies the Axiom Schema of Replacement.

For which α does V_α satisfy the Axiom Schema of Replacement? This is a delicate question. Consider the following formula:

(1) $$\varphi(n, y) : n \in \omega \wedge y = \omega + n.$$

By Claim 19, this formula is absolute for transitive models. Therefore,

(2) $$\langle V_{\omega+\omega}, \overline{\in}\rangle \models \forall n \in \omega \exists! y\, \varphi(n, y).$$

EXERCISE 24(G): (a) Convince yourself that

(3) $$\langle V_{\omega+\omega}, \overline{\in}\rangle \models \neg\exists z \forall n \in \omega \exists y \in z\, \varphi(n, y).$$

Hint: Use Corollary 6.

(b) Conclude that $\langle V_{\omega+\omega}, \overline{\in}\rangle$ is not a model of the Replacement Axiom Schema.

EXERCISE 25(PG): Show that neither $\langle V_{\omega\cdot\omega}, \overline{\in}\rangle$ nor any model of the form $\langle V_{\alpha+\omega\cdot 2}, \overline{\in}\rangle$ satisfies the Replacement Axiom Schema.

EXERCISE 26(R): Show that $\langle V_{\omega_1}, \overline{\in}\rangle$ does not satisfy the Replacement Axiom Schema.

Remark 20: Somewhat surprisingly, it is not possible to show that if $cf(\alpha) < \alpha$, then $\langle V_\alpha, \overline{\in}\rangle$ does not satisfy Replacement. In Exercise 25 we used the fact that the ordinals $\omega \cdot \omega$ and $\alpha + \omega \cdot 2$ are definable (the latter ordinal is definable with parameter $\alpha \in V_{\alpha+\omega\cdot 2}$). We will return to this issue in Chapter 24.

Remark 21: Note that we have shown that the Replacement Axiom Schema is not a consequence of the remaining axioms of ZF (all of them, and possibly even the Axiom of Choice, are satisfied in $\langle V_{\omega+\omega}, \overline{\in}\rangle$). By deleting the Replacement Axiom Schema from the axioms of ZF, one obtains a theory called *Zermelo's Set Theory* and abbreviated by Z. Adding the Axiom of Choice one obtains the Theory ZC. Since the most extensively studied mathematical objects (like, for example, the real line) can be coded as elements of $V_{\omega+\omega}$, an overwhelmingly large portion of mathematics can be formalized in ZC. On the other hand, ZC does not even imply the existence of the ordinal $\omega + \omega$.

Since getting a model of the form $\langle V_\alpha, \overline{\in}\rangle$ for all axioms of ZF seems a bit tricky, let us try our luck with the H_λ's. These models are much more inclined to satisfy Replacement, but there is a little problem with the Power Set Axiom. Let us investigate which H_λ's are closed under the power set operation. An initial ordinal λ is called a *strong limit cardinal* if $2^\kappa < \lambda$ for every $\kappa < \lambda$. Note that this definition makes sense even in ZF. [6] Moreover, note that $\aleph_0$ is a strong limit cardinal, and if GCH holds, every limit cardinal is strong limit.

EXERCISE 27(PG): Show that H_λ is closed under the power set operation if and only if λ is a strong limit cardinal.

Combining Lemmas 16 and 17 with Exercises 19–22, we obtain the following.

[6] If λ is a strong limit cardinal, then $\mathcal{P}(\kappa)$ can be wellordered for every $\kappa < \lambda$. Thus, in ZF alone one cannot prove the existence of an uncountable strong limit cardinal. But it is also relatively consistent with ZF that an uncountable strong limit cardinal exists while the full Axiom of Choice fails.

Theorem 22: *Suppose λ is an uncountable strong limit cardinal. Then H_λ is a model of* ZF.

EXERCISE 28(PG): Show that if λ is a strong limit cardinal, then $\langle H_\lambda, \overline{\in}\rangle \models$ (AC), even if the Axiom of Choice fails in **V**.

Apparently we have found a set model of ZF. But wait a minute. If $\lambda = \kappa^+$ for some κ, then $2^\kappa \geq \lambda$, and hence λ is not a strong limit cardinal. In other words, every strong limit cardinal is a limit cardinal. Therefore, if λ satisfies the assumptions of Theorem 22, then λ is weakly inaccessible. Even more is true. Let us call an uncountable strong limit cardinal *strongly inaccessible.* If GCH holds, then the notions of strongly inaccessible and weakly inaccessible cardinals coincide, but in general, it need not be the case that every weakly inaccessible cardinal is strongly inaccessible. Theorem 22 and Exercise 28 imply the following:

Corollary 23: ZF + "$\exists$ *a strongly inaccessible cardinal*" $\vdash$ CON(ZFC).

Proof: "CON(ZFC)" can be interpreted as "there exists a model for ZFC." □

Corollary 24: CON(ZF) $\not\to$ CON(ZF + *"there are no strongly inaccessible cardinals."*)

Proof: This follows immediately from Corollary 23 and Gödel's Second Incompleteness Theorem. However, the proof of Gödel's Theorem is beyond the scope of this book. Therefore, let us give an alternative, entirely self-contained argument.

EXERCISE 29(PG): Show that if λ is strongly inaccessible, then $H_\lambda = V_\lambda$.[7]

Now suppose all axioms of ZF hold in the universe **V**. If there are no strongly inaccessible cardinals in **V**, then there is nothing to prove. So suppose there are strongly inaccessible cardinals. By wellfoundedness of **ON** there exists a smallest one. The following lemma concludes the proof of Corollary 24.

Lemma 25: *Let λ be the smallest strongly inaccessible cardinal. Then $\langle V_\lambda, \overline{\in}\rangle \models$ "There are no strongly inaccessible cardinals."*

Proof: Let λ be as in the assumptions, and let $\varphi_{10}(\kappa)$ be the formula: "κ is a strongly inaccessible cardinal." We want to show that if $\kappa \in V_\lambda$, then $\langle V_\lambda, \overline{\in}\rangle \models \neg\varphi_{10}(\kappa)$. Of course, by the choice of λ, if $\kappa \in V_\lambda$, then κ is not a strongly inaccessible cardinal in **V**, so $\varphi_{10}(\kappa)$ fails (in **V**) for every $\kappa \in V_\lambda$. But the formula φ_{10} is not absolute for transitive models in general.[8] Fortunately, the following holds.

Sublemma 26: *Let δ be a limit ordinal. Then φ_{10} is absolute for V_δ.*

Proof: Let δ be as in the assumptions, and let $\kappa \in V_\delta$. Note that $\varphi_{10}(\kappa)$ can be written as $\varphi_1(\kappa) \wedge \varphi_{11}(\kappa) \wedge \varphi_{12}(\kappa)$, where:

---

[7] So, after all, we have found a model of ZF (even ZFC) of the form $\langle V_\alpha, \overline{\in}\rangle$!

[8] In order to show this, one needs more sophisticated tools than the ones presented here.

$\varphi_1(\kappa)$: $\kappa \in \mathbf{ON}$;
$\varphi_{11}(\kappa)$: "κ is a regular cardinal;"
$\varphi_{12}(\kappa)$: "κ is strong limit."

The idea of the proof is to show that failure of any one of the formulas $\varphi_1(\kappa)$, $\varphi_{11}(\kappa)$, $\varphi_{12}(\kappa)$ in $\mathbf{V}$ implies its failure in $\langle V_\delta, \overline{\in}\rangle$. The least troublesome of the three formulas is φ_1, since Claim 19 says that it is absolute for transitive models. The case of $\varphi_{12}(\kappa)$ is also relatively easy. Note that $\varphi_{12}(\kappa)$ can be written as: $\forall \alpha \in \kappa\, \varphi_{13}(\alpha, \kappa)$, where

$$\varphi_{13}(\alpha, \kappa): \ \exists f \exists p \exists \beta \in \kappa(\varphi_9(p, \beta, f) \wedge p = \mathcal{P}(\alpha)).$$

By Claim 19, φ_9 is absolute for transitive models. Moreover, by Exercise 18, the formula $y = \mathcal{P}(x)$ is absolute for models closed under the power set operation (in particular, for V_δ). Therefore, repeated application of the following claim shows that if φ_{12} fails in $\mathbf{V}$, then it fails in $\langle V_\delta, \overline{\in}\rangle$.

Claim 27: *Let $\mathbf{X}$ be a transitive class, let $y \in \mathbf{X}$, and let $\varphi(x)$ be a formula that is absolute for $\mathbf{X}$. Then the formulas $\exists x \in y\ \varphi(x)$ and $\forall x \in y\ \varphi(x)$ are absolute for $\mathbf{X}$. Moreover, if the formula $\exists x\, \varphi(x)$ fails in $\mathbf{V}$, then $\mathbf{X} \models \neg\exists x \varphi(x)$. Analogously, if the formula $\forall x\, \varphi(x)$ holds in $\mathbf{V}$, then $\mathbf{X} \models \forall x\, \varphi(x)$.*

EXERCISE 30(PG): (a) Prove Claim 27.
(b) Infer that if $\varphi_{12}(\kappa)$ fails in $\mathbf{V}$, then $\varphi_{13}(\alpha, \kappa)$ fails in $\langle V_\delta, \overline{\in}\rangle$ for all $\alpha \in \kappa$.
(c) Conclude that if $\varphi_{12}(\kappa)$ fails in $\mathbf{V}$, then $\varphi_{12}(\kappa)$ fails in $\langle V_\delta, \overline{\in}\rangle$.

Now consider the formula $\varphi_{11}(\kappa)$. It can be written as: $\forall f\, \varphi_{14}(f, \kappa)$, where

$\varphi_{14}(f, \kappa)$: "$\varphi_8(f) \wedge dom(f) \in \kappa \wedge rng(f) \subseteq \kappa \rightarrow \exists \beta \in \kappa (rng(f) \subseteq \beta)$."

EXERCISE 31(PG): Prove that the formula $\varphi_{14}(f, \kappa)$ is absolute for transitive models.

From Claim 27 and Exercise 31 we can infer the following.

Corollary 28: *Let $\mathbf{X}$ be a transitive class and assume $\kappa \in \mathbf{X}$. If $\varphi_{11}(\kappa)$ holds in $\mathbf{V}$, then $\mathbf{X} \models \varphi_{11}(\kappa)$.*

This corollary will come in handy at the end of this chapter, but for the present purpose it is useless, since we need its converse. The following observation helps out.

Claim 29: *Let $\alpha, \beta \in \mathbf{ON}$, assume f maps α into β, and let $\gamma = \max\{\alpha, \beta\}$. Then $f \in V_{\gamma+3}$.*

EXERCISE 32(G): Prove Claim 29.

By Claim 29, $\varphi_{11}(\kappa)$ is equivalent to the formula $\forall f \in V_{\kappa+3}\ \varphi_{14}(f, \kappa)$. Since $V_{\kappa+3} \in V_\delta$, we can infer from the first part of Claim 27 that if $\varphi_{11}(\kappa)$ fails in $\mathbf{V}$, then $\langle V_\delta, \overline{\in}\rangle \models \neg\varphi_{11}(\kappa)$. This concludes the proof of Sublemma 26, and hence of Lemma 25. □

Have we perhaps overlooked some interesting set models? In other words, are there transitive set models that are *not* of the form V_α? Perhaps yes, but such models cannot be closed under the power set operation.

EXERCISE 33(R): (a) Show that if $\mathbf{X}$ is a transitive class, then either $\mathbf{X} \cap \mathbf{ON} = \mathbf{ON}$ or $\mathbf{X} \cap \mathbf{ON} \in \mathbf{ON}$.

(b) Prove that the formula $\varphi(\beta, y)$: "$y = V_\beta$" is absolute for transitive classes that are closed under the power set operation.

(c) Conclude that if $\mathbf{X}$ is a transitive class closed under the power set operation, then either $\mathbf{X} = V_\alpha$ for some $\alpha \in \mathbf{ON}$, or $\mathbf{X} = \mathbf{V}$.

It follows from Exercise 33 that if we want to construct a proper, transitive class $\mathbf{X} \neq \mathbf{V}$ that is a model of ZF, then we must look for one that is not closed under the power set operation.

How can one construct such a class $\mathbf{X}$? If $\mathbf{X}$ satisfies the Power Set Axiom, then for every $x \in \mathbf{X}$ there exists a unique $z \in \mathbf{X}$ such that $\mathbf{X} \models \forall y(y \in z \leftrightarrow y \subseteq x)$. We shall denote this z by $\mathcal{P}^{\mathbf{X}}(x)$. By Claim 19(c), the formula $y \subseteq x$ is absolute for transitive models. Hence $\mathcal{P}^{\mathbf{X}}(x) \subseteq \mathcal{P}^{\mathbf{X}}(x)$. [9] There are various ways in which one could make $\mathcal{P}^{\mathbf{V}}(x)$ strictly smaller than $\mathcal{P}^{\mathbf{V}}(x)$. Let us explore just one possibility, that is, let us see what we get if we try to make $\mathcal{P}^{\mathbf{X}}(x)$ always *as small as possible.* We are going to build a model $\mathbf{L}$ step by step. At each step α, we will put the ordinal α itself into $\mathbf{L}$, and, given $x \in \mathbf{L}$ and $y \subseteq x$, we will put y into $\mathbf{L}$ only if we absolutely have to.

To be specific, let us consider $x = \omega$. When do we "absolutely have to" put a given $y \subseteq \omega$ into $\mathbf{L}$? Well, we want $\mathbf{L}$ to satisfy the Separation Axiom Schema. Thus, if there is a formula $\varphi(n)$ with one free variable such that y is the set of all $n \in \omega$ for which $\varphi(n)$ holds, then y must be in $\mathbf{L}$. As it turns out, *only* such $y \subseteq \omega$ need to be put into $\mathbf{L}$.

How come? What about the other axioms of ZF, like, for example, the Pairing Axiom? If $k, \ell \in \omega$, don't we also have to put $\{k, \ell\}$ into $\mathbf{L}$? Yes, but $\{k, \ell\} = \{n \in \omega : n = k \vee n = \ell\}$, so this is a definable subset of ω.

All right, but how can we get uncountably many subsets of ω this way? After all, the language L_S is countable, hence only countably many formulas can be formed in this language, and every formula defines at most one subset of ω. True, but in the Separation Axiom Schema we allow formulas that contain parameters. If a, b are different parameters, then the sets $\{n \in \omega : \varphi(n, a)\}$ and $\{n \in \omega : \varphi(n, b)\}$ may be different. So there is no reason why we couldn't have uncountably many definable (with parameters) subsets of ω.

But what if φ is not an absolute formula? This is a more serious objection than the previous two. Of course, if we want $\mathbf{L}$ to satisfy Separation, then $\mathbf{L}$ must contain every set of the form

$$\text{(C)} \qquad y = \{n \in \omega : \varphi_{/\mathbf{L}}(n)\}.$$

But what if, for example, φ is of the form $\forall z\, \psi(n, z)$? To keep matters simple, let us assume that ψ is absolute for transitive models. In order to decide whether or

[9]Beware of too obvious statements. The most fruitful method of constructing models of set theory, the forcing method (compare also Chapter 6) is based on the idea of making $\mathcal{P}^{\mathbf{X}}(x)$ *larger* than $\mathcal{P}^{\mathbf{V}}(x)$. If you want to learn how this is possible in spite of what we just said, we recommend K. Kunen's book *Set Theory,* North-Holland, 1980.

not to put y into **L**, we would have to know for which n the formula $\psi(n,z)$ holds for all $z \in \mathbf{L}$. So, in particular, we would have to check at some point whether $\psi(27,y)$ holds. Okay, but which y are we talking about here? The one that does contain 27, or the one that doesn't? Sounds familiar? We get the same problem of self-reference as in Russell's Paradox.

Fortunately, there is a trick that allows us to get around this problem. For any given set x, we can form the set *Def*(x) of all sets $y \subseteq x$ for which there are a formula $\varphi(v, a_0, \dots, a_{k-1})$ and parameters $a_0, \dots, a_{k-1} \in x$ such that

$$\text{(D)} \qquad y = \{z \in x : \langle x, \overline{\in} \rangle \models \varphi[z, a_0, \dots, a_{k-1}]\}.$$

Note that the definition (D) is immune to the kind of self-reference that can occur if we define y as in (C). Moreover, the following is true.

Theorem 30: *There exists a formula $\psi_0(x,y)$ of L_S such that ψ_0 is absolute for transitive models, and $\psi_0(x,y)$ holds iff $y = \mathrm{Def}(x)$.*

We omit the rather tedious proof of Theorem 30. Now we can recursively define the *constructible hierarchy* as follows:
$L_0 = \emptyset$;
$L_{\alpha+1} = Def(L_\alpha)$;
$L_\gamma = \bigcup_{\alpha<\gamma} L_\alpha \quad if \quad \gamma \in \mathbf{LIM}$.

Lemma 31: *Let $\alpha \in \mathbf{ON}$. Then:*
(i) *L_α is transitive;*
(ii) *$L_\alpha \subseteq V_\alpha$;*
(iii) *If $\alpha \leq \omega$, then $L_\alpha = V_\alpha$;*
(iv) *$\alpha \in L_{\alpha+1} \backslash L_\alpha$;*
(v) *$L_\alpha \subset L_\beta$ for every $\alpha < \beta$.*

EXERCISE 34(PG): Prove Lemma 31.

Now let $\mathbf{L} = \bigcup_{\alpha \in \mathbf{ON}} L_\alpha$. The class **L** is called the *constructible universe;* the sets in **L** are called the *constructible sets.* It follows from Theorem 30 and Theorem 8.5 that there exists a formula $\psi_1(\alpha, z)$ such that $\psi_1(\alpha, z)$ holds if and only if $\alpha \in \mathbf{ON}$ and $z = L_\alpha$. Now let $\psi_2(x)$ be the formula:

$$\exists \alpha \exists z \, \psi_1(\alpha, z) \wedge x \in z.$$

Note that ψ_2 is a representation of **L**, i.e., $\mathbf{L} = \{x : \psi_2(x)\}$. Finally, let $\mathbf{V} = \mathbf{L}$ be the sentence: $\forall x \, \psi_2(x)$. Now we are ready to state the main theorem of this chapter.

Theorem 32 (Gödel's): (a) $\mathbf{L} \models \mathbf{V} = \mathbf{L}$.
(b) $\mathbf{L} \models$ ZF.
(c) $\mathbf{L} \models$ *"There exists a set-like wellorder of the universe* **V**.*"*
(d) $\mathbf{L} \models$ GCH.

Point (c) implies the strong version $(\mathrm{AC})^+$ of the Axiom of Choice introduced at the end of Section 9.1.

Corollary 33: $\mathbf{L} \models$ *"There exists a functional class* **F** *with domain* $\mathbf{V} \backslash \{\emptyset\}$ *such that* $\mathbf{F}(x) \in x$ *for every* $x \in \mathbf{V} \backslash \{\emptyset\}$.*"*

EXERCISE 35(G): Derive Corollary 33 from Theorem 32(c).

Only the axioms of ZF are needed for the construction of **L** and the proofs of Theorem 32 and Corollary 33. Thus, starting with a universe where the axioms of ZF are supposed to hold, one can find a model for ZFC + GCH. This yields the following.

Corollary 34 (ZF): $CON(\mathrm{ZF}) \to CON(\mathrm{ZFC} + \mathrm{GCH})$.

Sketch of the proof of Theorem 32: One can show that the formulas ψ_0, ψ_1, and ψ_2 mentioned above are absolute for **L**. This yields (a). Somewhat surprisingly, the only really difficult part of the proof of (b) is the Separation Axiom Schema. Suppose $x \in \mathbf{L}$, and $\varphi(z, a_0, \dots, a_{k-1})$ is a formula with one free variable z. Let $\alpha \in \mathbf{ON}$ be such that $x, a_0, \dots a_{k-1} \in L_\alpha$. At every stage $\beta \geq \alpha$ of the construction of **L** we dutifully put the set $y_\beta = \{z \in x : \langle L_\beta, \overline{\in}\rangle \models \varphi[z, a_0, \dots, a_{k-1}]\}$ into **L**. But the relevant instance of the Separation Axiom Schema requires something apparently stronger, namely that the set $y_\infty = \{z \in x : \mathbf{L} \models \varphi(z, a_0, \dots, a_{k-1})\}$ is in **L**. Fortunately, the following lemma implies that, indeed, $y_\infty \in \mathbf{L}$.

Lemma 35: *Let $\alpha \in \mathbf{ON}$ and $x, a_0, \dots, a_{k-1} \in L_\alpha$, and let $\varphi(z, a_0, \dots, a_{k-1})$ be a formula of L_S. Then there exists a $\beta > \alpha$ such that for every $z \in x$:*

$$\langle L_\beta, \overline{\in}\rangle \models \varphi[z, a_0, \dots, a_{k-1}] \quad \textit{iff} \quad \mathbf{L} \models \varphi[z, a_0, \dots, a_{k-1}]$$

Lemma 35 is an instance of a more general theorem, called the *Reflection Principle.* The Reflection Principle will be discussed in Chapter 24.

EXERCISE 36(R): Prove that **L** satisfies the remaining axioms of ZF.

How could one construct a wellorder of **L**? The very definition of **L** suggests that one should do this by recursion. To begin with, let $(\varphi_n)_{n\in\omega}$ be an enumeration of all formulas of L_S that are not sentences. Now suppose L_α has been wellordered by a relation $\leq_\alpha$. This generates a lexicographic order relation $\leq_\alpha^k$ on $(L_\alpha)^k$. Note that $\leq_\alpha^k$ is also a wellorder relation. Now suppose φ_n has one free variable x and parameters $a_0, \dots, a_{k(n)-1}$. Given $a_0, \dots, a_{k(n)-1} \in L_\alpha$, let

$$y_n(a_0, \dots, a_{k(n)-1}) = \{x \in L_\alpha : \langle L_\alpha, \overline{\in}\rangle \models \varphi_n[x, a_0, \dots, a_{k(n)-1}]\}.$$

For $y \in L_{\alpha+1} \backslash L_\alpha$, we let $n(y) = \min\{n : y = y_n(a_0, \dots, a_{k(n)-1})\}$, and denote by $\overline{a}(y)$ the $\leq_\alpha^{k(n(y))}$-minimal $k(n(y))$-tuple $\langle a_0, \dots, a_{k(n)-1}\rangle$ such that $y = y_{n(y)}(a_0, \dots, a_{k(n(y))-1})$. Now we define a relation $\leq_{\alpha+1}$ on $L_{\alpha+1}$ as follows: $y \leq_{\alpha+1} z$ iff

(i) $y, z \in L_\alpha \wedge y \leq_\alpha z$ or
(ii) $y \in L_\alpha \wedge z \notin L_\alpha$ or
(iii) $y, z \in L_{\alpha+1} \backslash L_\alpha \wedge n(y) < n(z)$ or
(iv) $y, z \in L_{\alpha+1} \backslash L_\alpha \wedge n(y) = n(z) \wedge \overline{a}(y) \leq_\alpha^k \overline{a}(z)$.

It is a routine, but somewhat tedious exercise to check that $\leq_{\alpha+1}$ is a wellorder relation on $L_{\alpha+1}$ that extends $\leq_\alpha$. Now we can inductively define $\leq_\beta$ for every β. At limit stages γ we simply take $\leq_\gamma = \bigcup_{\alpha<\gamma} \leq_\alpha$. This way we get a wellorder $\mathbf{W} = \bigcup_{\beta\in\mathbf{ON}} \leq_\beta$ of the whole class **L**. By (ii), this wellorder is set-like.

Now comes the bad news: Our informal description of **W** amounts to a recursive construction of a proper class. As we have seen earlier in this text, even recursive constructions of sets require the Axiom of Choice. Recursive constructions of proper classes require even something stronger, namely $(\mathrm{AC})^+$. But we promised you that Theorem 32 would be provable in ZF. Therefore, in any complete proof of Theorem 32 one has to check that the class $\mathbf{W}$ is definable, i.e., that $\mathbf{W} = \{\langle x, y\rangle : \psi_3(x, y)\}$ for some formula ψ_3. Such a formula ψ_3 does exist, but we will not construct it here.

Finally, let us now sketch the proof of (d). It is based on the following observation.

Claim 36: (a) $|Def(x)| = |x|$ *for every infinite set* x.
(b) $|L_\alpha| = |\alpha|$ *for every infinite ordinal* α.

EXERCISE 37(G): Prove Claim 36.

It follows that the constructible hierarchy grows at a much more leisurly pace than the cumulative hierarchy. While for example $|V_{\omega+2}|$ is already as big as $2^{2^{\aleph_0}}$, L_α is countable for every countable ordinal. By Claim 36, the following lemma suffices for the proof of (d).

Lemma 37: *Assume* $\mathbf{V} = \mathbf{L}$, *and let* κ *be a cardinal. Then* $\mathcal{P}(\kappa) \subseteq L_{\kappa^+}$.

Sketch of the proof: We need the following fact.

Claim 38: *There exists a finite subset* T *of the theory* ZF + $(\mathbf{V} = \mathbf{L})$ *such that whenever* x *is transitive and* $\langle x, \overline{\in}\rangle \models T$, *then there exists* $\gamma \in \mathbf{ON}$ *such that* $x = L_\gamma$.

The proof of Claim 38 is similar to Exercise 33, except that one uses absoluteness of the formula "$y = L_\beta$" instead of absoluteness of the formula "$y = \mathcal{P}(x)$."

Now suppose $\mathbf{V} = \mathbf{L}$ holds, let κ be as in the assumptions of Lemma 37, and suppose $y \in \mathcal{P}(\kappa)$. Fix an ordinal $\alpha > \kappa$ such that $y \in L_\alpha$. Since the theory T of Claim 38 is finite, we can form the conjuction φ of all sentences in T. By Lemma 35, there exists a $\beta \geq \alpha$ such that $\langle L_\beta, \overline{\in}\rangle \models \varphi$, or equivalently, $\langle L_\beta, \overline{\in}\rangle \models T$. By a result that will be proved in Chapter 24, there exists a set $N \subseteq L_\beta$ such that:

(1) $\kappa \subseteq N$ and $y \in N$;

(2) $|N| = \kappa$;

(3) for every $k \in \omega$, for every formula $\psi(a_0, \dots, a_{k-1})$ of L_S with k free variables, and for every k-tuple $\langle a_0, \dots, a_{k-1}\rangle$ of elements of N we have

$$\langle N, \overline{\in}\rangle \models \psi[a_0, \dots, a_{k-1}] \text{ iff } \langle L_\beta, \overline{\in}\rangle \models \psi[a_0, \dots, a_{k-1}].$$[10]

In particular, by (3), $\langle N, \overline{\in}\rangle \models T$. However, N is not in general transitive. Now consider the Mostowski collapse M of N. This M is transitive. Moreover, the collapsing function f is an isomorphism of the models $\langle N, \overline{\in}\rangle$ and $\langle M, \overline{\in}\rangle$. So, in particular, $\langle M, \overline{\in}\rangle \models T$. By Claim 38, $M = L_\gamma$ for some $\gamma \in \mathbf{ON}$. By (2), $\gamma < \kappa^+$.

EXERCISE 38(PG): Let M, N be as above, and let $f : N \to M$ be the collapsing function. Show that $f(\alpha) = \alpha$ for every $\alpha \in \kappa$, and conclude that $f(y) = y$. *Hint:* Use (1).

[10]An N that satisfies (3) is called an *elementary submodel* of L_β.

Exercise 38 implies that $y \in L_\gamma$. Thus, by Lemma 31(v), $y \in L_{\kappa^+}$, and we have proved Lemma 37. □

The class **L** is the first example of an *inner model.* Inner models are constructed by identifying a certain property P of sets ("being constructible" in our case) and reinterpreting the notion of "set" as "set with property P."

Inner models are a powerful tool to obtain so-called *equiconsistency results.* Two theories S and T are called *equiconsistent* if ZF $\vdash CON(S) \leftrightarrow CON(T)$. In this terminology, we can say that Corollary 34 establishes one direction of the equiconsistency of ZF and ZFC + GCH (the other direction is obvious).

In Chapter 11, we introduced the notion of a large cardinal and explained why one should, if possible, avoid using them in proofs (see the discussion preceding and following Fact 11.21). But suppose φ is a sentence of L_S such that you were able to prove CON(ZFC + "There exists a measurable cardinal") $\rightarrow CON$(ZFC + φ). You obtained a consistency result, but you paid a price in the form of making a slightly dubious assumption. It makes sense to ask whether the price was right. Could you have gotten by with a somewhat weaker assumption, like CON(ZFC + "There exists a strongly inaccessible cardinal"), or even just plain CON(ZFC)? This is where inner model theory comes in. In many cases, it is possible to assume that the axioms of ZFC together with your sentence φ hold in the universe **V**, and to construct under these assumptions an inner model **X** such that **X** $\models$ "There exists a measurable cardinal." The existence of such an inner model proves that CON(ZFC + φ) $\rightarrow CON$(ZFC + "There exists a measurable cardinal"), and it follows that the two theories ZFC + φ and ZFC + "There exists a measurable cardinal" are equiconsistent. In other words, the price you paid in your consistency proof of ZFC + φ was right.

Let us conclude this chapter with an example of an equiconsistency proof. Since every strongly inaccessible cardinal is weakly inaccessible, it is a rather trivial observation that CON(ZFC + "There exists a strongly inaccessible cardinal") implies CON(ZFC + "There exists a weakly inaccessible cardinal"). But weakly inaccessible cardinals need not be strongly inaccessible. Nevertheless, the *consistency strength* of the existence of a weakly inaccessible cardinal is the same as that of the existence of a strongly inaccessible cardinal.

Theorem 39: *CON(ZFC + "There exists a weakly inaccessible cardinal") $\rightarrow$ CON(ZFC + "There exists a strongly inaccessible cardinal").*

Proof: Suppose the universe **V** satisfies ZFC + "There exists a weakly inaccessible cardinal," and let κ be weakly inaccessible. Then $\kappa \in$ **L**, and by Corollary 28, **L** $\models$ "κ is a regular cardinal." The following exercise shows that we also have **L** $\models$ "κ is a limit cardinal."

EXERCISE 39(PG): Show that for every $\lambda < \kappa$, if **L** $\models$ "λ is a cardinal," then **L** $\models \lambda^+ < \kappa$.

Therefore, **L** $\models$ "κ is a weakly inaccessible cardinal." Since **L** satisfies the GCH, we can conclude that **L** $\models$ "κ is a strongly inaccessible cardinal." □

Mathographical Remarks

Kurt Gödel first announced Theorem 32 in two short articles that appeared in 1938 and 1939. The complete proof was published in 1940. The monograph *Constructibility,* by Keith J. Devlin, Springer, 1984, gives a thorough presentation of the constructible universe and its various properties. Devlin's book is also a good starting point for those readers who wish to learn more about inner model theory.

Subject Index

Index of Notation

Greek letters (and $\aleph$) are listed under the first letter of their English name. Special symbols ($\emptyset$, $=$, etc.) and operations ($\alpha + \beta$, κ^+, etc.) are grouped separately at the end.